Disaster Management and Private Sectors

Disaster Management and Private Sectors

Editor

Kiran Deshpande

Disaster Management and Private Sectors

Edited by **Kiran Deshpande**

Printed in 2017

ISBN: 978-1-68117-126-5

Library of Congress Control Number: 2015936548

© 2016 by
SCITUS Academics LLC,
616, Corporate Way, Suite 2, 4766,
Valley Cottage, NY 10989

www.scitusacademics.com

Contents

Preface

Disaster Management is the coordination and integration of all activities necessary to build, sustain and improve the capability to prepare for, protect against, respond to and recover from threatening or actual natural or human-induced disasters. It is a multijurisdictional, multi-sectoral, multi-disciplinary and multi-resource initiative. Therefore, it is vital that the Federal, State and Local Governments, Civil Society Organisations (CSOs) and the private sector discharge their respective roles and responsibilities and complement each other in achieving shared goals of disaster management. The Emergency Services Sector (ESS) is a system of preparedness, response, and recovery elements that form the Nation's first line of defense for preventing and mitigating the risk from physical and cyber attacks, and manmade and natural disasters. The sector consists of emergency services facilities and associated systems, trained and tested personnel, detailed plans and procedures, redundant systems, and mutual-aid agreements that provide life safety and security services across the Nation via a first-responder community comprised of Federal, State, local, tribal, territorial, and private sector partners.

Editor

The Importance of Social Value in the Evaluation of Web Services in the Public Sector

Sunghwan Lee[1], Sangun Park[2], and Wooju Kim[1]

[1]Department of Information and Industrial Engineering, Yonsei University, 50 Yonsei-ro, Seodaemun-gu, Seoul 120-749, Republic of Korea

[2]Division of Business Administration, Kyonggi University, Gwanggyosan-ro, Yeongtong-gu, Suwon-si, Gyeonggi-do 443-760, Republic of Korea

ABSTRACT

Wireless sensor networks (WSNs) are widely used in many different fields. Even in the public sector, various services using WSN are offered. One of the key issues is how to control and manage heterogeneous devices of WSN devices. Devices Profile for Web Services (DPWS), a standard of Web services, has been adopted to solve the problems of interoperability between WSN devices. In order to evaluate WSN services in the public sector, this paper presents a method to evaluate Web services, a base technology of sensor network services. This paper presents a value analysis methodology assessing tangible and intangible benefits of Web services in the public sector. We classify stakeholders of Web services as a government, citizens, and agents (businesses) and selected the metrics for each stakeholder's benefit. After that, we determine the weight of each metric through AHP. The result shows that social value was the most important benefit in the construction of Web services in the public sector. We expect that the main contribution of this paper is the development of a value assessment framework that reflects the unique characteristics of Web services in the public sector.

INTRODUCTION

Wireless sensor network (WSN) is used in many various fields recently [1, 2]. As a result, various studies are carrying out and, in particular, it became necessary to control and manage heterogeneous devices of WSN. Web services have been adopted to solve the problems of interoperability between devices, and a variety of services using WSN with Web services were developed [3–7]. In the private sector as well as in the public sector, services including waste management, disaster management, and traffic information were offered by using WSN. Most of services offered by the smart city use Web services [3, 4, 6, 8, 9]. As Web services became widely used in terms of both aspects of technology and service, there is a growing need to consider proper use of Web services, when evaluating WSN in the public sector.

The smart city concept is used in many countries as a strategy to implement e-government and a lot of cities such as Vienna, Amsterdam, Dubai, and Yokohama provide various services for the smart city [10–12]. E-government becomes an enabler for countries to achieve national competitiveness for the sake of development and wellbeing of their citizens [13–15]. In that sense, Web services became one of the catalysts that could promote better communication between government agencies and citizens [16–19]. Many countries adapted Web services for their e-government IT systems [10, 20]. Moreover,

Web services provide standards for enterprises and governments to integrate an application infrastructure cost effectively. The standards also help to compose new service-oriented businesses and make third party software marketplaces.

However, research about evaluation of the proper use and construction of Web services in the public sector is insufficient [21, 22]. Therefore, we suggest an evaluation method for Web services in the public sector. The most important issue of our research is the fact that Web services in the public sector have different characteristics about stakeholders compared to the private sector.

It is a noticeable characteristic that all of the stakeholders (government, agency, business, and citizen) can get benefits, respectively, through Web services in the public sector. Therefore, we develop metrics for each stakeholder's benefit. We expect that the social value would be the most important value in our framework while it was not significantly considered in most previous studies that focused on the private sector.

RELATED WORK

Wireless Sensor Networks and Web Services

A wireless sensor network (WSN) means a device of sensors autonomously monitoring physical or environmental conditions distributed in a certain space, such as temperature, pressure, and sound, and cooperatively passing their data to a main location through the network. WSN is a group of wireless sensor nodes. Wireless sensor node is a tiny independent device with the capability of wireless communication, sensing, and computing [23, 24]. A numerical increment of devices requires managing device interaction and interoperability simple way. Web services handled interoperability issue. Web services are the international standard technology supporting the interoperability between different kinds of operating systems, networks, and development programming language [25]. The first proposal version of the Devices Profile for Web Services (DPWS) was released in 2004 [7]. The DPWS specify a minimum set of constraints on the implementation on devices with limited resources.

It enables secure Web services messaging and dynamic Web services discovery and description. The SIRENA (Service Infrastructure for Real-Time Embedded Networked Applications) project developed the first framework adapting DPWS [26]. DPWS are currently an OASIS standard [7], which is chosen as a suitable subset of Web services protocols for machine-to-machine (M2M) communication. As Web services technologies have widespread use in the WSN, it is important to assess the web services.

Wireless Sensor Networks in the Public Sector

WSNs have an important impact on multiple fields such as environmental change detection, health care monitoring, and supply chain management. In particular, in the public sector, waste management, disaster management, facilities management, health care, educational services, cultural tourism services, rental services, traffic information, illegal parking management, and structure monitoring are utilizing WSN.

Such services put all together can be defined as the smart city. The aim of smart city is to make better connection of important city infrastructure components (city administration, education, healthcare, public safety, real estate, transportation, and utilities) and services [27]. In Harrison et al.'s study, a smart city is defined as an instrumented, interconnected, and intelligent city [8]. Instrumentation enables acquiring and integrating live real-world data by utilizing systems of sensors, meters, and software in IT systems. Interconnection means the integration of information acquired from instrumented systems through public and private networks in the city and the communication of such information across multiple processes including exchange of public services by various city agencies. Intelligent refers to analysis, optimization, and making decisions based on interconnected information for operational efficiency and improvement of quality of citizen's life.

Review on Public Service Value Measurement Methodologies

Recently, a considerable amount of efforts has been undertaken to develop concepts and methodologies to capture the value creation

The Importance of Social Value in the Evaluation of Web Services...

5

of Information and Communication Technology (ICT) projects in the public sector. The following list provides some examples:

- balanced e-government index [28];
- demand and value assessment methodology [29];
- government performance framework [30];
- performance reference model [31, 32];
- public sector value model [33];
- value measuring methodology [34, 35];
- value of investment methodology [36, 37].

While these methodologies have various approaches to value measurement, most of them attempted to apply private sector metrics, such as traditional Return on Investment (ROI) measures, to the public domain.

Compared to other methodologies, the Value Measuring Methodology (VMM) model is noticeable due to its diversity of evaluation value factors, including the social value considered in this study. Therefore, we selected the VMM model for the base evaluation framework in this study. VMM [34, 35] is a methodology for measuring the values of e-services developed by the Social Security Administration and the General Services Administration in the USA.

Theoretical Framework: Value Measuring Methodology

Value Measuring Methodology (VMM) is an evaluation model that helps decision-makers weigh tangible and intangible values when making a decision or observing benefits. Other methods to calculate the Return on Investment (ROI) have been used for many years, but there was no widely formal method to provide grounds for decisions on the basis of intangible values. For decision-makers, it was difficult to keep a balance between intangible benefits and costs, especially in case where corporations make a plan of long-term investments, and governments and nonprofit institutions that are primarily interested in intangible values make a plan of using funds within limited budget. VMM includes the perspectives of stakeholders concerned with the initiative, including direct users and government partners. VMM shows the gap between current tools and satisfies the necessity for a new different analysis methodology of planning, proposal, investment, management, and assessment.

The application of VMM starts from developing a structure of values, such as costs, risks, tangible benefits, and intangible benefits, and then giving importance to each factor in the structure. If an agreement on the relevant importance of each type of values is made, it enables decision-makers to review alternatives, consent decisions in an objective and repeatable way, and compare respective values in a project. The quantitative application of the VMM permits analysis of the contribution in total to a certain value across a range of projects. VMM was developed through the project Evaluation on Values of Electric Service in 2001 jointly conducted by Social Security Administration's Office (SSA) and General Services Administration (GSA) of the United States federal government. Booz Allen Hamilton and Harvard University's John F. Kennedy School of Government published the results from this project in a report, Building a methodology for measuring the value of e-service. In October 2002, the Best Practices Committee of CIO (The Federal Chief Information Officer) Council published a guidebook of VMM titled Value Measuring Methodology: How-To-Guide [35]. Other countries and nongovernment organizations adapted the documents from the guidebook. Accordingly, the VMM model was developed to evaluate both the quantitative and qualitative values of the government's research and development program.

DESIGN OF EVALUATION MODEL

Stakeholders of Web Service in the Public Sector

The aim of Web services in the public sector is to provide benefits to government agencies and citizens. Therefore, the stakeholders of Web services in the public sector are classified as the government (government ministries initiating the Web services project), the government agencies (agencies conducting the Web services project), business (Web services system development vender), and citizens (see Figure 1).

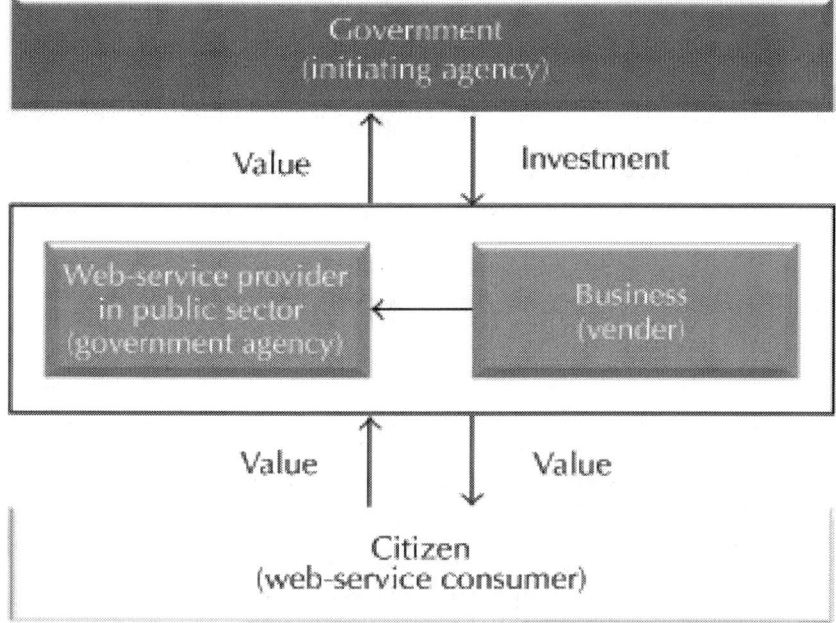

Figure 1: Stakeholders of Web services in the public sector.

When the government initiates a project order with a Web services provider and IT service provider, values occur at the time that the Web services provider offers services to IT service providers (e.g., cost saving). When the Web services provider offers its services to citizens, citizens can get benefits (e.g., citizens can save time by using the Web services).

Definition of Value Structure

As discussed in the previous sections, benefits from Web services in the public sector differ depending on stakeholders, and so existing information system evaluation models are not appropriate for the public sector. After comparing evaluation models, the VMM model was selected for the basic evaluation framework because it covers the widest range of stakeholders considered.

The value structure of VMM was designed to evaluate the following five value factors: (i) direct user value–value to a direct user of the information system; (ii) social value–value to society as a whole, generated by the information system; (iii) government financial value–financial value to the government; (iv) government operational value–value associated with the improvement of government operations; and (v) strategic value–value to the government in terms of policy implementation. Table 1 shows the definitions and corresponding stakeholders of the values.

Table 1: Values and stakeholders of Web services in the public sector

Values	Definition	Stakeholders
Direct user value	Benefits directly realized by users	Citizen
		Government agency
Social value	Public benefit related to society as a whole	Society as a whole/ the public
		Citizen
Government operational value	Improvement in the initiating agency's business process	Government/ government agency
Government financial value	Economic benefits for affiliated agency/ initiating agency	Government/ government agency
Strategic value	Contribution to the government's strategic goal	Government

Benefits of Using Web Services

Web services technology provides many potential benefits. In order to evaluate public Web services, we conducted a literature review for identifying the benefits obtained by introducing Web services. After project completion, the finished project should be assessed to determine whether or not the stakeholders actually benefited from the project. This study examined the benefits of introduction of Web services by reviewing numerous papers. A summary of work by Chen [38, 39], Chen et al. [40], Ciganek et al. [41, 42], Hailstone and Perry [43], Lee [44], and Wilkes [45] is presented as shown in Table 2.

Table 2: Benefits from Web services adoption

Values	Benefits
Increasing Benefit	Providing new service
	Coordination
	Service usability improvement
	Efficient use of resources
	Improving work efficiency
	Collaboration (intra-/ extracollaboration)
Saving cost	S/W cost
	H/W cost
	Labor cost (development/ operation)

PRIORITIZATION OF VALUES AND MEASURES

Although our framework considers exactly necessary factors for the public sector, there are priorities of factors in decision-making processes. In our proposed framework, this prioritization process was done by using an Analytic Hierarchy Process (AHP), which establishes rankings by comparing each value factor against the others. AHP is a decision-making technique developed by Thomas L. Saaty, a professor at the University of Pittsburgh in the United States [46]. In case where the evaluating criteria for the decision goal are complex and multiple, AHP is used to decompose the structure of the decision into multilevel hierarch and to establish priorities of factors through pairwise comparison of such factors.

This study used the following 4 steps of AHP generally applied.

(Step 1) Structure the decision hierarchy.

(Step 2) Collect judgments data through pairwise comparison of factors.

(Step 3) Measure relative weights.

(Step 4) Calculate relative total weights.

We established a decision hierarchy to determine values of web services according to AHP as in Figure 2. The decision goal is the value of web services and 5 value factors were selected as the evaluation criteria in the 1st level. Each value factor has 3 or 4 subfactors as the evaluation criteria in the 2nd level.

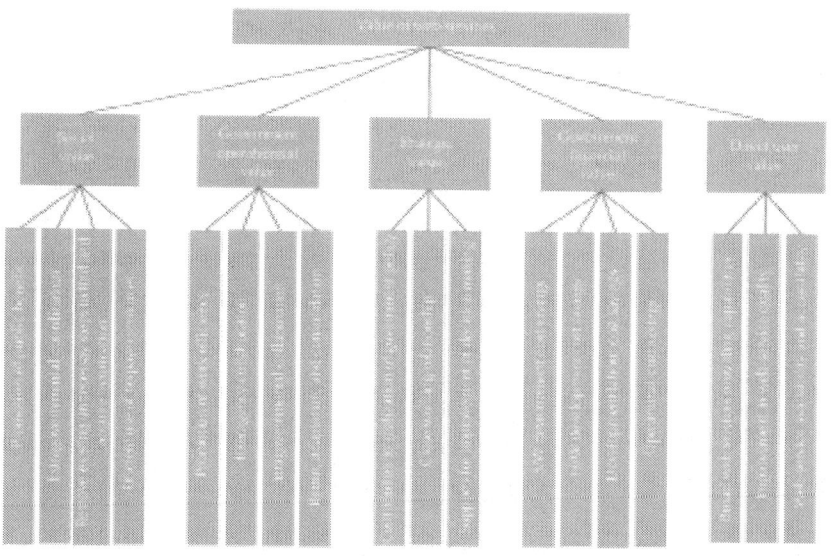

Figure 2: Value tree of Web Services.

After structuring the decision hierarchy, we developed AHP survey questionnaires and organized evaluation questionnaires for pairwise comparison based on 9-point scale. For AHP survey, we selected 22 professionals and the professional group participating in this AHP survey is composed of 5 professors, 14 graduate students whose major is diagnosis of IT project, 1 public e-service provider, and 2 researchers of the government department related to the public e-service. After the design of value factors and metrics, AHP analysis was performed by a group of professionals to determine the relative importance of individual values and details of metrics. In this process, this study used only the results below 0.2 of consistency ratio in order to judge the logical consistency of individual responses. Thereby, we calculated the weights for each factor reflecting expertise of professionals. The survey described in Table 3 shows that the stakeholders set a high value on the social value relatively to others.

Table 3: Weighted value and measures of the web services in the public sector evaluation model

Value	Measures	Weights
Social value 26.96	Promotion of public benefit	10.66
	Extragovernmental coordination	6.66
	Removing existing unnecessary steps to find and acquire information	5.46
	Efficient use of taxpayer resources	4.18
Government operational value 24.81	Promotion of work efficiency	8.89
	Interagency collaboration	7.01
	Intragovernmental collaboration	5.56
	Reuse, adaptation, and consolidation	3.35
Strategic value 17.8	Contribution to realization of government policy	13.04
	Close working relationship	1.62
	Supports for improvement of decision-making	3.14
Government financial value 16.9	S/W development cost savings	4.46
	H/W development cost savings	3.37
	Development labor cost savings	3.87
	Operational cost savings	5.20
Direct user value 14.93	Broad Web-service providing capabilities	2.71
	Improvement in Web-service quality	6.66
	Web-service availability and accessibility	5.56
Total value weights		100

CONCLUSIONS

In order to evaluate WSN services in the public sector, this paper presented a method to evaluate Web services, a base technology of sensor network services. We proposed an evaluation model for

Web services in the public sector by adapting the Value Measuring Methodology as our base framework. We suggested evaluation model reflecting benefits of stakeholders of Web services in the public sector and weighted each metric by utilizing AHP analysis. As the result, we found that social value has the highest weight. This implies that social value should be preferentially considered when designing and assessing Web services in the public sector. We expect that the findings in this study can be used as reference to assess Web services in the public sector. Moreover, it suggested that the social value should be considered important in the evaluation of Web services. This model we designed has been applied to evaluation in practice and used for policy data.

REFERENCES

1.   M. J. Kim and J. H. Shim, "A beacon scheduling for mesh topology in wireless sensor networks," The Journal of Society for e-Business Studies, vol. 15, no. 4, pp. 49–58, 2010.

2.   K.-C. Shin, "A robust biometric-based user authentication protocol in wireless sensor network environment," The Journal of Society for e-Business Studies, vol. 18, no. 3, pp. 107–123, 2013.

3.   M. S. Familiar, J. F. Martínez, and L. López, "Pervasive smart spaces and environments: a service-oriented middleware architecture for wireless ad hoc and sensor networks," International Journal of Distributed Sensor Networks, vol. 2012, Article ID 725190, 11 pages, 2012.

4.   N. Glombitza, D. Pfisterer, and S. Fischer, "Integrating wireless sensor networks into web service-based business processes," in Proceedings of the 4th International Workshop on Middleware Tools, Services and Run-Time Support for Sensor Networks (MidSens '09), pp. 25–30, December 2009.

5.   R. Kyusakov, J. Eliasson, J. Delsing, J. van Deventer, and J. Gustafsson, "Integration of wireless sensor and actuator nodes with IT infrastructure using service-oriented architecture," IEEE Transactions on Industrial Informatics, vol. 9, no. 1, pp. 43–51, 2013.

6.   G. Moritz, F. Golatowski, C. Lerche, and D. Timmermann, "Beyond 6LoWPAN: Web services in wireless sensor networks," IEEE Transactions on Industrial Informatics, vol. 9, no. 4, pp. 1795–1805, 2013.

7.   OASIS Standard, Devices Profile for Web Services Version 1.1, 2009.

8.   C. Harrison, B. Eckman, R. Hamilton et al., "Foundations for smarter cities," IBM Journal of Research and Development, vol. 54, no. 4, pp. 1–16, 2010.

9.   G. Moritz, E. Zeeb, F. Golatowski, D. Timmermann, and R. Stoll, "Web services to improve interoperability of home healthcare devices," in Proceedings of the 3rd International Conference on Pervasive Computing Technologies for Healthcare, 2009. PervasiveHealth, pp. 1–4, IEEE, April 2009.

10. H. Chourabi, T. Nam, S. Walker et al., "Understanding smart cities: an integrative framework," inProceedings of the 45th Hawaii International Conference on System Sciences (HICSS '12), pp. 2289–2297, January 2012.

11. A. S. City, Introduction Amsterdam Smart City, Municipality of Amsterdam, 2009, http://amsterdamsmartcity.com/assets/media/factsheet_launch_Amsterdam_Smart_City_June_3rd.pdf.

12. S. Yamamoto, S. Matsumoto, and M. Nakamura, "Using cloud technologies for large-scale house data in smart city," in Proceedings of the 4th IEEE International Conference on Cloud Computing Technology and Science (CloudCom '12), pp. 141–148, December 2012.

13. G. Aichholzer, "Scenarios of e-government in 2010 and implications for strategy design," Electronic Journal of e-Government, vol. 2, no. 1, pp. 1–10, 2004.

14. S. Shan, L. Wang, J. Wang, Y. Hao, and F. Hua, "Research on e-Government evaluation model based on the principal component analysis," Information Technology and Management, vol. 12, no. 2, pp. 173–185, 2011.

15. S. K. Sharma, "Assessing e-government implementations," Electronic Government, vol. 1, no. 2, pp. 198–212, 2004.

16. R. S. Baraka and S. M. Madoukh, "A conceptual SOA-based framework for e-Government central database," in Proceedings

of the International Conference on Computer, Information and Telecommunication Systems (CITS '12), pp. 1–5, May 2012.

17. J. Gamper and N. Augsten, "The role of web services in digital government," in Electronic Government, vol. 2739 of Lecture Notes in Computer Science, pp. 161–166, Springer, Berlin, Germany, 2003.

18. S. K. Goudos, N. Loutas, V. Peristeras, and K. Tarabanis, "Public administration domain ontology for a semantic web services Egovernment framework," in Proceedings of the IEEE International Conference on Services Computing (SCC '07), pp. 270–277, IEEE, Salt Lake City, Utah, USA, July 2007.

19. B. Medjahed, A. Rezgui, A. Bouguettaya, and M. Ouzzani, "Infrastructure for e-government Web service," IEEE Internet Computing, vol. 7, no. 1, pp. 58–65, 2003.

20. Z. Ebrahim and Z. Irani, "E-government adoption: architecture and barriers," Business Process Management Journal, vol. 11, no. 5, pp. 589–611, 2005.

21. F. Al Shahwan, K. Moessner, and F. Carrez, "Evaluation of distributed SOAP and RESTful mobile web services," International Journal on Advances in Networks and Services, vol. 3, no. 3-4, pp. 447–461, 2011.

22. R. Mizouni, M. A. Serhani, R. Dssouli, A. Benharref, and I. Taleb, "Performance evaluation of mobile web services," in Proceedings of the 9th European Conference on Web Services (ECOWS '11), pp. 184–191, September 2011.

23. W. Dargie and C. Poellabauer, Fundamentals of Wireless Sensor Networks: Theory and Practice, John Wiley & Sons, Chichester, UK, 2010.

24. K. Sohraby, D. Minoli, and T. Znati, Wireless Sensor Networks: Technology, Protocols, and Applications, John Wiley & Sons, 2007.

25. Web Services Architecture Working Group W3C, Web Services Architecture, Web Services Architecture Working Group W3C, 2004.

26. H. Bohn, A. Bobek, and F. Golatowski, "SIRENA—service infrastructure for real-time embedded networked devices: a service oriented framework for different domains," in Proceedings

of the International Conference on Networking, International Conference on Systems and International Conference on Mobile Communications and Learning Technologies (ICN/ICONS/MCL '06), p. 43, Morne, Mauritius, April 2006.

27. D. Washburn, S. Balaouras, and L. E. Nelson, "Helping CIOs understand 'Smart City' initiatives,"Growth, vol. 17, 2009.

28. Bertelsmann Foundation and Booz Allen & Hamilton, Balanced E-Government Index—Messung von E-Government/ E-Democracy: Hintergrundinformationen zur Methode, Bertelsmann Foundation and Booz Allen & Hamilton, 2002.

29. AGIMO, Demand and Value Assessment Methodology, AGIMO, 2004.

30. A. di Maio, New Performance Framework Measures Public Value of IT, Gartner, Stamford, Conn, USA, 2003.

31. FEAPMO, The Performance Reference Model Version 1.0: A Standardized Approach to IT Performance—Volume I: Version 1.0 Release Document, 2003.

32. FEAPMO, The Performance Reference Model Version 1.0: A Standardized Approach to IT Performance—Volume II: How to Use the PRM, 2003.

33. V. Jupp and M. P. Younger, "A value model for the public sector," Outlook, vol. 1, pp. 16–21, 2004.

34. CIO Council, The Value Measuring Methodology: Highlights, 2001.

35. CIO Council, Value Measuring Methodology: How to Guide, CIO Council, Framingham, Mass, USA, 2002.

36. T. Trigon, "Cost & benefit analysis of TESTA," Tech. Rep., 2003.

37. T. Trigon, "DA Value of Investment Method—Version 2.1.," 2003.

38. M. Chen, "Factors affecting the adoption and diffusion of XML and web services standards for e-business systems," International Journal of Human Computer Studies, vol. 58, no. 3, pp. 259–279, 2003.

39. M. Chen, "An analysis of the driving forces for Web services adoption," Information Systems and e-Business Management, vol. 3, no. 3, pp. 265–279, 2005.

40. A. N. K. Chen, S. Sen, and B. B. M. Shao, "Strategies for effective Web services adoption for dynamic e-businesses," Decision Support Systems, vol. 42, no. 2, pp. 789–809, 2006.

41. A. P. Ciganek, M. N. Haines, and W. Haseman, "Challenges of adopting web services: experiences from the financial industry," in Proceedings of the 38th Annual Hawaii International Conference on System Sciences (HICSS '05), p. 168b, IEEE, January 2005.

42. A. P. Ciganek, M. N. Haines, and W. Haseman, "Horizontal and vertical factors influencing the adoption of Web services," in Proceedings of the 39th Annual Hawaii International Conference on System Sciences (HICSS '06), vol. 6, p. 109, January 2006.

43. R. Hailstone and R. Perry, "IBM and the strategic potential of web services: assessing the customer experience," IDC, 2002.

44. S. Lee, "Benefit analysis, due to the introduction of web services," in Proceedings of the Korea Society of MIS Fall Conference, pp. 383–388, 2007.

45. L. Wilkes, ROI—The Costs and Benefits of Web Services and Service Oriented Architecture, 2004.

46. T. L. Saaty, "Decision making with the analytic hierarchy process," International Journal of Services Sciences, vol. 1, no. 1, pp. 83–98, 2008.

Chapter 2

Disaster Management Discourse in Bangladesh: A Shift from Post-Event Response to the Preparedness and Mitigation Approach through Institutional Partnerships

C. Emdad Haque[1] and M. Salim Uddin[1]

[1]Natural Resources Institute, University of Manitoba, Winnipeg, Canada

INTRODUCTION

The discourse of disaster management has undergone significant changes in recent decades and their effects have been profoundly felt in the developing world, particularly in terms of reduction in the loss of human lives. In this chapter, we concentrate on the evolution of disaster management approaches in Bangladesh and the method of their implementation by mobilizing institutions as a case in the developing world. The geographical location of Bangladesh in South Asia, at the confluence of three large river systems – the Brahmaputra, the Ganges, and the Meghna – and north of the Bay of Bengal, renders it one of the most vulnerable places to floods and cyclones. Human-induced climate change exacerbates the problem, with its already manifested effects and the predicted rise in sea level of 0.3 m to 0.5 m by 2050 [1, 2, 3]. Climate models have revealed that the effects of climate change are not only affecting individual countries, but resulting in increased climate variations at regional levels [4]. Bangladesh, as part of South Asia, is likely to experience more variations in climate regimes, as well as more extreme weather events.

Bangladesh is the most densely populated country in the world, except city states, with more than 1,000 people per sq km [5]. Agriculture, which provides a quarter of the gross domestic product (GDP), depends largely on timely monsoon rainfall and regularity in seasonal fluctuations. In the period 1970–2004, about 0.7 million people lost their lives due to natural disasters, and economic losses totaled about US $5.5 billion [6, 7, 8, 9]. The cyclone of 1970, in the coastal areas of what was then East Pakistan, alone claimed over half a million lives. Again, the 1991 and 2007 cyclones caused the loss of about 149,000 and 3,406 people respectively. In recent years, the frequency of extreme floods has increased, as has the corresponding economic loss. The flood in July 2004 was the most devastating – with an economic loss of about US $2.2 billion [5]. In terms of GDP, this loss was less than what the world's poorest countries faced during the 1985–99 disasters – a loss of 13.4% of combined GDP [10]. But the loss in Bangladesh amounted to an immense step backwards in development efforts. The floods in 2007 inundated about 36% of the total area in 57 out of 64 districts [11] and affected at least 4.5 million people [12].

Because of the extreme vulnerability of the people in general and of the economy to natural disasters, various regimes of the government of Bangladesh (pre- and post-independence) have developed an institutional infrastructure to deal with natural hazards and their potential losses [13, 14]. Traditionally, the disaster management approach in the country has been to respond to disaster in the aftermath of the events. Nonetheless, the ever-increasing human and economic costs have raised serious questions about such approaches. Also, flood-disaster management has relied heavily on structural engineering and post-flood relief operations. Overall, such a post-facto approach has failed to effectively deal with the problems of disaster loss.

In recent years, there has been a shift to recognize the critical roles of non-structural measures as well as pre-disaster mitigation and preparedness. These initiatives recognize the roles of different stakeholders. For example, the Disaster Management Act of 1998 acknowledges the capacity of affected populations [15]: "An event, natural or man-made, sudden or progressive, that seriously disrupts the functioning of a society, causing … such severity that the affected community has to respond by taking exceptional measures and is on a scale that exceeds the ability of the affected people to cope with using only its own resources." Disaster management warrants more than relief and recovery: it should be part of development planning, without which the loss of life and property is likely to intensify. It is recognized that institutional partnerships can be effective when they involve all stakeholders – government, local communities, NGO/CBOs, media, the private sector, academia, neighboring countries, and donor communities.

In this study, we examine the extent and effectiveness of institutional partnership from the perspective of a shift from a managerial approach to an approach using participatory, collective decision-making and resource-sharing to manage disaster risk. Since community members are the direct and most seriously affected victims, effective and sustainable partnership requires a change from a partnership approach based on equality to a focus on the community [16]. Our central concerns are to assess who decides, at what level, and how. There has been only very limited analysis of the shifting approaches and of how institutions at different levels are presently functioning in Bangladesh. Is this mechanism based on partnership, with collective decision-making? Is a culture of working together on a national cause such as disaster management evolving? How functional are these elaborate institutional mechanisms? What is the role of the private sector or the knowledge stakeholders? How can an effective partnership be built into disaster management? These are questions we examine in this chapter.

METHODOLOGY

Our research examines whether the elaborate institutional mechanisms of disaster management in Bangladesh reflect the partnership of stakeholders. For this purpose, social science policy analysis seemed useful. Obtaining reliable quantitative data on activities of both government and NGOs is a chronic problem in Bangladesh; our research therefore applied qualitative, rather than quantitative, tools and techniques.

Both primary and secondary data sources were used for empirical investigation and policy analysis. To collect the primary data from local communities, we applied a case study approach which employed participatory rural appraisal tools, such as focus group discussions (FGDs), household interviews and key informant interviews, in two coastal communities of Bangladesh severely affected by Cyclone Sidr. We analyzed data procured from a total of 162 households distributed across eight villages, which were randomly selected to conduct the interviews and FGDs. We received 100% response from the targeted households for household interview. A random sampling procedure was followed to select households from the complete list of households in the selected villages for interview. Households were proportionately selected according to the size of the villages, and household heads were interviewed. Eight FGDs were carried out by administering a semi-structured questionnaire to each village.

For policy analysis, we relied chiefly on secondary data, which were supplemented by primary data. Because Bangladesh has experienced numerous devastating cyclones, as well as long-lasting floods that have caused immense suffering to people and damage to properties in recent decades, we relied on secondary data on the relevant disaster response and management policies. Official documents from the government and donors, study reports from NGOs and other organizations, journal articles, newspaper clippings, TV reports and documentaries, and internet resources from reliable and responsible sources provided additional information for our analysis. To ensure openness in discussing sensitive issues, we used informal discussions with stakeholders at different levels. Through personal contacts and over the internet we collected reports and documents from government agencies, NGOs, and donors in Bangladesh, such as the Bangladesh Disaster Preparedness Centre (BDPC), the Disaster Forum, the Disaster Management Bureau, the Sustainable Development Resource Centre, and the United Nations Development Program (UNDP).

SHIFTING APPROACHES IN DISASTER MANAGEMENT

In recent decades, government agencies, non-governmental organizations (NGOs) and local communities in Bangladesh have undertaken various measures to mitigate the impacts of natural disasters, including floods and cyclones, on the people, economy and society. The concept of developing national preparedness, as opposed to post-event response, to disasters like floods and cyclones evolved after the floods of 1988 and the devastating cyclone of 1991. The main argument behind this shift was that if people were well prepared for frequent disasters they would minimize their impacts, resulting in a reduced need for relief and rehabilitation. It was also strongly felt by the public institutions that if disaster preparedness could be integrated in the socio-economic development process at household, community, regional and national levels, it would build the long-term capacity of the community to mitigate risk and vulnerability to disasters [17]. The aim of the shift also included changing disaster management approaches and measures from structural engineering interventions to the social dimensions and community partnerships.

The stated significant change in emergency and disaster management approaches demanded institutional restructuring in the governance portfolios. Consequently, the government of Bangladesh, led by the Ministry of Disaster Management and Relief, has undertaken various steps in the form of policy, strategy and programs considering the concept of disaster management through mitigation, preparedness, recovery and rehabilitation. The government established the Disaster

Management Bureau (DMB) under the Ministry of Disaster Management in 1993 to promote disaster prevention, mitigation and preparedness; to provide guidelines; and to organize training and awareness for the concerned people and stakeholders to mitigate the impacts of disasters. Currently, the DMB has focused on risk reduction through community mobilization, capacity-building and linking risk reduction with the socio-economic development of the poor and vulnerable groups and with developing the DMB's partnership with other government agencies, NGOs and international organizations.

Alongside the development thinkers, international development partners (such as UNDP, DFID, Oxfam GB, USAID, Care International, Caritas) and local NGOs which are concerned with, and are experienced in, disaster management in Bangladesh, the government has promoted the approach of capacity-building and disaster preparedness at all levels. A call for institutional partnerships therefore stemmed from both the government as well as non-governmental and civil society organizations. A few key policy planners and senior government officials have also favored this new thinking and reflected this through the renaming of the Ministry of Relief and Rehabilitation to the Ministry of Disaster Management and Relief. In 1993, the Ministry of Disaster Management then established the Disaster Management Bureau (DMB) and the government set up a national council and various committees at national, district, *upazila* (sub-district) and union (local council) levels, for overall disaster management preparedness [17]. The implications of this institutional restructuring were manifold, and evolved through a sequence of placing increasing emphasis on institutional partnership and community-based disaster management (CBDM).

Institutional Restructuring to Reflect a Shift in Disaster Management

In order to manage the consequences of natural disasters, formal public policymaking institutions in Bangladesh have formulated a well-developed mechanism (Figure 1) at national and field levels. The factors that led to such a development can be explained as follows:

- the severity of the consequent casualties has led to motivations at local, national and international levels to address the issue;

- the recurrent disasters created serious development setbacks: loss in the production and infrastructure sectors set back the affected regions and the country; and

- in order to attract external investment, the minimization of disaster risks and vulnerabilities warranted intervention at the policy level.

In Bangladesh, at the national level, four high-profile bodies were established for the multi-sectoral coordination of emergencies associated with environmental disasters as well as disaster management in general: the National Disaster Management Council (NDMC), headed by the prime minister; the Inter-Ministerial Disaster Management Coordination Committee (IMDMCC), led by the minister of food and disaster management; the National Disaster Management Advisory Committee (NDMAC), headed by a specialist nominated by the prime minister; and a Parliamentary Standing Committee on Disaster Management to supervise national policies and programs. The common missions of these bodies have been to provide policy and management guidance and the macro-coordination of activities, particularly relief and rehabilitation.

Presently, the lead actor in disaster management is the Ministry of Disaster Management and Relief (MoDMR) until 2002. It has the role of inter-ministerial planning of disaster management and coordination and of responding in the event of a disaster. Under the MoDMR, there are two line agencies, the Disaster Management Bureau (DMB) and the Directorate of Relief and Rehabilitation (DRR). The DMB is a small professional unit at the national level that performs specialist functions, working with district and *upazila* (sub-district) administrations and line ministries under the overall guidance of the IMDMCC. It is a catalyst for planning, for arranging public education, and for organizing the systematic training of government officers and other personnel from the national down to the union (local council) or community level. The DRR manages the post-disaster provision of relief and rehabilitation. At present, it leads risk reduction at the local community level.

Among all the other ministries, the Ministry of Water Resources (MoWR) plays a vital role in flood management. It is involved in the planning of water resources, including planning for water-related natural disasters such as cyclone protection, flood proofing, riverbank erosion control and drought management, although the mitigation of disasters remains beyond their mandate. The Flood Forecasting and Warning Center (FFWC), under the Bangladesh Water Development Board (BWDB) of the MoWR, plays an important role in providing early warning about impending floods to the agencies involved.

In the areas of both cyclone and flood hazards, the Bangladesh Red Crescent Society (BRCS) and various donor agencies play important roles. The Cyclone Preparedness Program (CPP) was established in 1972 following the devastating cyclone of 1970, under an agreement between the BRCS and the government of Bangladesh, with an aim to undertake effective cyclone preparedness measures in the coastal areas. CPP, under the BRCS, has a joint management structure, with two committees, *viz.* a 7-member Policy Committee headed by the Minister of MoFDM, and a 15-member Implementation Board led by the Secretary of the MoFDM. Now the CPP has about 33,120 trained volunteers, including 5,520 women [18].

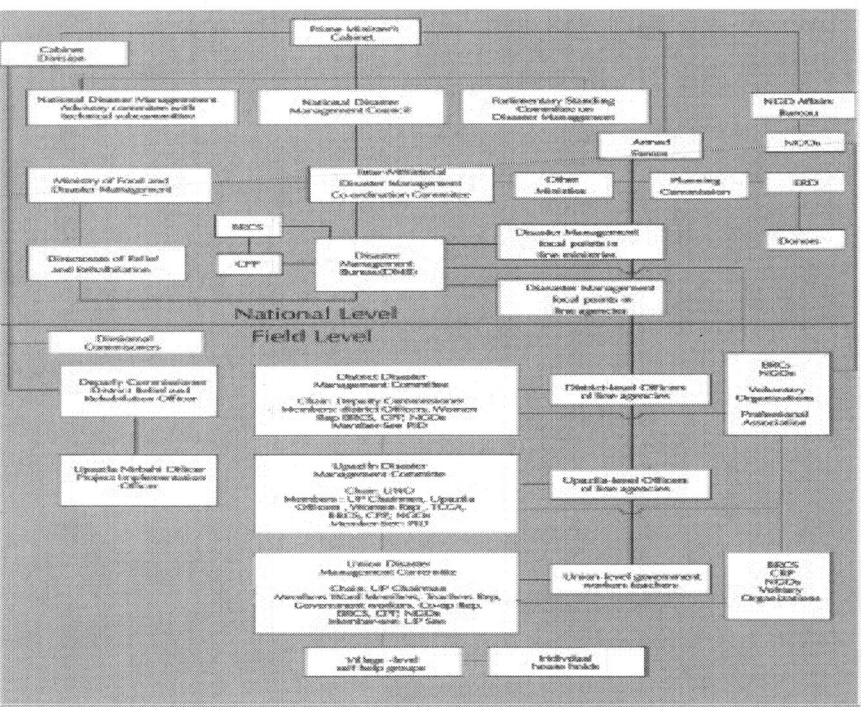

Figure 1: Organizational structure and institutional arrangements for disaster management at the national level and field level.

Besides, the government has a "standing order" for natural disasters (mainly for floods and cyclones), which was last updated in August 1999. The standing orders are followed by all ministries, divisions/departments and government agencies during normal times, precautionary and warning stages, the disaster stage and the post-disaster stage.

The National Water Management Plan also underlines the importance of implementing effective non-structural measures to reduce the impact of floods and erosion. Thus, as opposed to the structural measures against floods (like dams, river embankments, and flood control and drainage projects) and riverbank erosion control projects (like the building of hard points, canalization and revetment), the recent policies and plans have recognized the importance of participatory planning that focuses on sustaining people's livelihoods.

At the field level (Figure 1), disaster management and related mechanisms start with the district administrations covering all 64 districts of Bangladesh. The District Disaster Management Committee (DDMC) is chaired by the deputy commissioner, the chief civil administrator of the district. The members of the committee include departmental officers and NGO, BRCS and CPP and women's representatives. Likewise, below the district level, there are the *upazila*, union and village tiers of the disaster management committees. These local-level committees include representatives from almost all relevant interest groups in society (Figure 1). An examination of how these committees function appears in succeeding sections.

Increasing Roles and Responsibilities of NGOs

The Disaster Management Bureau (DMB) has been assigned the role of coordinating the activities of NGOs. The NGOs constitute a vibrant sector in Bangladesh, and have been acclaimed worldwide. NGOs and CBOs are actively involved, among others, in disaster management, micro-credits, family planning, and human rights protection. As a matter of fact, the advent of NGO activities in Bangladesh owes its origin to the rehabilitation works immediately after the devastating war of independence in 1971. Currently, about a quarter of foreign assistance to Bangladesh is channeled through the NGOs. Therefore, their contributions, particularly to the social service sector and the mobilization of the poor, are quite prominent. This has been acclaimed by the international community. NGOs like the Grameen Bank and Bangladesh Rural Advancement Committee (BRAC) have extended their development and disaster management programs at the international level as well.

NGOs such as CARE-Bangladesh, OXFAM-Bangladesh, Action Aid, Intermediate Technology Development Group-Bangladesh, Bangladesh Disaster Preparedness Center (BDPC) and Disaster Forum are particularly involved in various pre-, during and post-disaster activities. Pre-disaster activities include advocacy, public education campaigns and training programs for personnel involved in disaster management from the national down to the union or local community level. NGOs also are active in emergency evacuation and in taking people to shelters. The post-disaster activities include offering new micro-credits or rescheduling their loan payment programs for rehabilitation.

Developments in the Institutional Framework: Introduction to the Comprehensive Disaster Management Plan (CDMP)

Besides the Cyclone Preparedness Program (CPP) and the standing orders, the government of Bangladesh adopted a Corporate Plan (2005-2009) called "Comprehensive Disaster Management: A Framework for Action." The US $15 million Comprehensive Disaster Management Plan (CDMP,Table 1) was funded by DFID and UNDP. It aimed "to reduce the level of community vulnerability to natural and human-induced hazards and risks to manageable and humanitarian levels." This program was supposed to be implemented through a "program-based approach" that encompassed all aspects of risk management. The approach comprehended a transition from a single agency response and relief system to a holistic strategy involving the entire development planning process of the government. CDMP Phase II was launched in 2010 to institutionalize the adoption of risk reduction approaches, and to channel support through government and velopment partners, civil society and NGOs into a people-oriented disaster management and risk reduction partnership. The project period will be 2010-20114.

Table 1: Sub-programs, outputs, target area/group and implementing agencies of CDMP. Adapted from [19]

Strategic Directions	Sub-programs	Target Area/ Group	Key Outputs	Responsible Agencies
Raising the level of expertise of the Disaster Management Systems	Capacity-building	MDMR and Implementing Agency	1. PPDU established and effectively executing its key functions. 2. New MDMR allocation of business and organogram reflecting broader responsibilities in disaster risk management 3. Professional skill enhancement program developed and implemented 4. Professional training institutionalized 5. Phase II program identified	MDMR/ UNOPS
Mainstreaming Disaster Risk Management Programming	Partnership Development	National, District and Upazila level officials	1. High level advocacy program established and implemented 2. Review of the development project appraisal processes and integration of disaster risk management 3. Training for national, district and upazila officials implemented	MDMR/ DMB/ NGO MDMR/ DMB

Strengthening Community Institutional Mechanisms	Community Empowerment	Union, Ward and Community levels	1. Inventory of existing programs developed and gaps identified	UNOPS
				DRR
			2. Community risk management programs based on formal hazard analysis	UNOPS
			3. The Local Risk Reduction Fund is supporting community risk reduction efforts	
Expand Preparedness Programs across a broad range of hazards	Research, Information and Management	Dhaka and selected cities	1. Urban search and rescue pilot for Dhaka fire services based on earthquake threat	MDMR/Fire Service
			2. Establishing an integrated approach to climate change risk management at national and local levels	MoEF/ DoE
Operationalizing Response Systems	Response Management	Whole country	1. Upgrading the capacity in information management during normal and emergency periods	MDMR
			2. Regional networks strengthened	
			3. Timely deployment of resources	
			4. Operational response capacities strengthened	

The Corporate Plan (2005-09; 2010-14) acknowledged the need for pre-disaster mitigation and preparedness of the people as opposed to the earlier concepts of responding after a disaster had taken place. Priority was accorded to focus on community-level preparedness, response, recovery and rehabilitation. Programs to train people living in disaster-prone areas were emphasized to improve their capability to cope with natural disasters. The Corporate Plan emphasized a series of broad-based strategies: First, disaster management involved the *management of both risks and consequences* of disasters, which included prevention, emergency response and post-disaster recovery. Second,*community involvement* was a major focus for preparedness programs to protect lives and properties. The involvement of local government bodies was an essential part of the strategy. Self-reliance was the key for preparedness, response and recovery. Third, *non-structural mitigation measures*, such as community disaster preparedness, training, advocacy and public awareness, were given a high priority; this required the integration of structural mitigation with non-structural measures.

The strategic focus of the CDMP was to lay the foundation for the shift in principle from a post-disaster relief and response strategy towards a comprehensive risk minimization culture that encouraged disaster-resilience initiatives. This approach was to be realized through a series of interconnected strategic directives:

- Raising the level of expertise of the disaster management systems,
- Mainstreaming disaster risk management programming,
- Strengthening community institutional mechanisms,
- Expanding preparedness programs across a broad range of hazards, and
- Putting the response systems into operation.

Based on these directives, the major sub-programs of CDMP included: (1) Capacity-building, (2) Partnership Development, (3) Community Empowerment, (4) Research and Information Management, and (5) Response Management. Under the sub-program of *Partnership Development*, the government actively sought to achieve a multi-agency approach that encompassed the institutions of the government, NGOs and private sector in a collaborative strategy for the alleviation of disaster-induced poverty. This enhanced coordination and information-sharing among the various actors and thus maximized the efficacy of resource use for effective risk reduction. Under the *Community Empowerment* sub-program, the government planned to further consolidate the empowerment process by expanding the program and by realizing community capacity-building through awareness and skill development and by expanding disaster management studies within the school system and staff training academies.

Besides these, disaster risk reduction was incorporated as a component into the Poverty Reduction Strategy Paper (PRSP) of Bangladesh as Annex-9 of Disaster Vulnerability and Risk Management [14]. The preparation of the PRSP, funded by the World Bank, acknowledged a holistic approach to disaster management.

SHIFT FROM RELIEF AND RESPONSE TO DISASTER RISK MANAGEMENT

In the last few decades, disasters have typically been viewed by the public institutions as numerous individual extreme events, and the responses included top-down-oriented government policies and efforts by local and international relief agencies that did not take into simultaneous account the social and economic implications and causes of these events. With the significant advancement in the understanding of the natural processes that underlie the hazardous events, a more technocratic paradigm came into existence which conceded that the only way to deal with disasters was by the public policy application of geophysical and engineering knowledge and the associated interventions. These approaches treated disasters as exceptional or "abnormal" events, not related to the ongoing social and developmental processes. Gradually, with recognition of the fact that these are not "natural events" per se, but directly linked with social structures and their dynamics [20], this structural engineering, technocratic approach shifted to an emphasis on preparedness measures, such as stockpiling of relief goods, preparedness plans and a growing role for relief agencies such as the Red Crescent [21]. This evolution of public policy approach from "relief and response" to "risk management" has begun to influence the way disaster management programs are planned and financed. Initiatives have been aimed more and more at reducing social and economic vulnerability and at investing in long-term mitigation activities.

Community-Based Disaster Management (CBDM)

Recognizing the need for vulnerability reduction for effective disaster management, the failures of a top-down management approach have become evident. This approach has been unsuccessful in addressing the needs of vulnerable communities. A better understanding of disasters and losses also brings to light the fact that the increased occurrence of disasters and disaster-related loss has been due to the exponential increase in the occurrence of small- and medium-scale disasters. As a result, numerous scholars and stakeholders feel that it is important to adopt a new strategy that directly involves vulnerable people in the planning and implementation of mitigation, preparedness, response, and recovery measures. This bottom-up approach has received wide acceptance because it considers communities as the best judges of their own vulnerability and capable of making the best decisions regarding their well-being. The search for the newer approach led to the formulation of the Community-Based Disaster Management (CBDM) strategy.

The aim of CBDM is to reduce vulnerabilities and to strengthen people's capacity to deal with hazards and cope with disasters. A thorough assessment of a community's exposure to hazards and an analysis of their specific vulnerabilities and capacities is the basis for activities, projects and programs that can reduce disaster risks. Because a community is involved in the whole process, their real and felt needs, as well as their inherent resources, are considered. Therefore, there is a greater likelihood that problems will be addressed with appropriate interventions. People's participation is not only focused on processes but also on the contents. It is anticipated that the local community should be able to gain directly from improved disaster risk management. This in turn will contribute to a progression toward safer conditions and to the improved security of livelihoods and sustainable development. This underlines the point that the local community is not only the primary actor but also the beneficiary of the risk reduction and development process [21].

The implementation of CBDM requires consideration of many essential features. Following Yodmani (2001), [21], the primary ones could be identified as:

- The local community has a central role in long-term and short-term disaster management and therefore the focus of attention in disaster management must be on the local community.

- Disaster risk or vulnerability reduction is the foundation of CBDM; the primary content of disaster management activities revolves around reducing vulnerable conditions and the root causes of vulnerability. The primary strategy for vulnerability reduction is by increasing a community's capacities and their resources, and by improving and strengthening coping strategies.

- Disaster management must also establish linkages to the development process as disasters are viewed as unmanaged development risks and unresolved problems of the development process. CBDM should lead to a general improvement of the quality of life of the vast majority of the poor people and of the natural environment.

- CBDM contributes to people's empowerment – to possess physical safety; to have more access to, and control of, resources; to participate in decision-making which affects their lives; to enjoy the benefits of a healthy environment.

- As community is a key resource in disaster risk reduction, their role and interests must be recognized. The community is the key actor as well as the primary beneficiary of disaster risk reduction. Within the community, priority attention is given to the conditions of the most vulnerable, as well as to their mobilization in the disaster risk reduction. The community must directly participate in the whole process of disaster risk management -- from situational analysis to planning and to implementation.

- A multi-sectoral and multi-disciplinary and trans-disciplinary approach must be applied. CBDM brings together the multitude of community stakeholders for disaster risk reduction, as well as to expand their resource base. The local community-level institutions link up with the intermediate and national levels and even up to the international level to address the complexity of vulnerability issues. A wide range of approaches to disaster risk reduction is employed.

- The CBDM is an involving and dynamic framework, and therefore its implementation must be monitored, evaluated and adapted to incorporate newer elements. Lessons learned from practice continue to build into the theory of CBDM. The sharing of experiences, methodologies and tools by communities and CBDM practitioners continues to enrich practice.

IS THE PRESENT FRAMEWORK BASED ON A PARTNERSHIP APPROACH?

As is evident from the institutional structure explained above, Bangladesh has developed quite an elaborate framework and disaster preparedness and response mechanism. Moreover, some policy and plan pronouncements in the recent past indicate that the government has begun to adopt an approach to disaster management that includes both risks and consequences. Some progress has been made in enhancing the disaster management capacities during the last decades. After the experiences of the devastating 1988 floods and 1991 cyclone, the concept of *disaster management* was introduced in place of *disaster control*. The ministry was renamed the Ministry of Disaster Management and Relief (MoDMR) in 1993 and then again renamed the Ministry of Food and Disaster Management (MoFDM) in 2002. After the formation a new government in 2008, this name of this ministry went back to its previous title as the Ministry of Disaster Management and Relief (MoDMR).

The primary function associated with disaster management is outlined in the government's Rules of Business which are undertaken by the DMB and the DRR. The Rules of Business have been revised to reflect the current MoDMR approach of comprehensive, community-based vulnerability reduction and risk management. The result is that though there has been a declining trend in loss of lives and property, particularly from cyclone disasters, flood damage has tended to rise because of the large spatial extent of floods, their increased frequency and the expanding economy.

Government documents and the NGO literature indicate that there is a wide recognition that effective disaster response at the local level is not possible by government agencies alone and that the cost of management needs to be shared by all stakeholders. Still, the major lacuna in the institutional framework continues to be a lack of functioning partnerships among the stakeholders. The massive flood of July 2004 showed that there were no partnerships functioning and there was little coordination. The Local Consultative Group (LCG) concluded that massive shortcomings existed in the forecasting, preparedness and coordinated response to the crisis [22]. As a result, the NGOs conducted relief and rehabilitation efforts largely without government directives and coordination. Initially, the government appeared confident to deal with the post-disaster recovery singlehandedly. When things were getting worse, it made a flash appeal on 17 August 2004 through the UNDP, Dhaka, for international assistance. Another report argues about the handling of 1998 floods, indicating that "limited evidence of government coordination was found in the recovery phase" [23]. Save the Children (USA) also proclaimed that "there was a general lack of coordination among actors" [24]. In the wake of the latest cyclone, Sidr in 2007, BBC reported, "Plenty of agencies, but not enough aid - too little, too late," and further quoted a professional working in an affected area, "The reason why these people are not receiving enough help is because there is no coordination between the government and aid agencies" [25].

A striking example of poor management and coordination is the following case. Southkhali village in Shoronkhola *upazila* of Bagerhat district was one the worst hit areas in Sidr. During a visit immediately after the event to the area, the Indian foreign minister pledged his country's intention to build all the houses in this and the surrounding villages. From then onwards, nominal government initiative was taken to give shelters to the affected people in this area, and a virtual official ban was put into effect on others, including NGOs and aid agencies, to build houses for the affected people. The pledged Indian support did not come in due time and even 100 days after the event people in this area were forced to live under the sky [26]. Perhaps this unfortunate decision arose from the lack of international/bilateral coordination, bureaucracy on both sides in Bangladesh and India, a lack of understanding of the gravity of not giving shelter to victims in time, or from the unnecessary exercise of power on the administration's part, even when in distress.

Empirical Investigation of Cyclone Sidr Victims at the Local Level: Vulnerability of the Poor

It is worth noting that the recent initiatives on community-based disaster risk management became subject to stern criticisms because of their general inadequacy in addressing the vulnerability of the poor to natural hazards and socioeconomic shocks. CBDM programs that aim at prevention and mitigation are few in low- and middle-income countries like Bangladesh and they are poorly funded and insignificant when compared to the financial capital spent by donors and development banks on humanitarian assistance, relief and post-disaster reconstruction. Another weakness of such initiatives was that they were often taken up in the formal sector of the economy, and therein bypassed the poor and the most vulnerable sections of society. As Maskrey (p. 86) points out, "in the year or so between the occurrence of a disaster and approved national reconstruction plans, many vulnerable communities revert to coping with risk, often in the same or worse conditions than before the disaster actually struck" [27]. Therefore, in the current paradigm of risk management approaches, there is more room than ever before for addressing the issues of risk reduction for the poor. This is also consonant with the paradigm shift in mainstream development practice, which is now characterized by a focus on good governance, accountability and greater emphasis on bottom-up approaches [21]. In light of the above perspective, the case of Cyclone Sidr can be examined.

Bangladesh has experienced several catastrophic environmental disasters during the last two decades; among these events, the April 1991 [28] and November 2007 major cyclones were the most catastrophic in terms of both physical and human dimensions [29]. Cyclone Sidr struck the coast of Bangladesh on 15 November 2007 and was the most powerful mega-cyclone to impact Bangladesh since 1991. However, the death toll (officially, 3,406 lives were lost) caused by Cyclone Sidr was significantly lower than comparable cyclones in previous years due to the improved warning system and evacuation. Nonetheless, the damage to crops and infrastructure was considerable across 30 districts, 200 *upazilas* and 1,950 unions. In total, more than 55,000 people were injured by the Cyclone Sidr event. The Joint Damage, Loss, and Needs Assessment (JDNLA) committee estimated that the total damage and losses caused by the cyclone were more than US \$1.7 billion [30].

Our investigation along the coastal plains of Bangladesh revealed that the geographical location and patterns of settlement associated with low income populations were the most important determinants of vulnerability to tropical cyclones and related storm surges. Coastal and island people observed that the new, isolated, single-unit settlements were the most severely impacted by the cyclone and storm surge forces. The settlements near the coast and those which were in linear patterns along the coastal embankment suffered the most. This type of settlement, which was more susceptible to cyclonic storm surges, was inhabited primarily by the poor of the coastal zones. The fragmentation of families and the building of new settlements also contributed to high cyclone disaster loss. The soil of the new settlements was less cohesive, and in most cases, the properties had few or only very small trees. The houses of the linear settlements along the coastal and island embankment were made of straw, bamboo and other locally sourced materials. Apart from high winds and storm surges, these houses were also vulnerable to breaches of the embankment that occur in major cyclonic storms. In our survey, we registered that most of the houses constructed using straw, bamboo, jute stalk, and corrugated iron sheets alongside of the coastal embankment had faced the sea. Such positioning of the housing structures also made them more susceptible to severe damage by the cyclones. In contrast, houses located in the interior mainland were usually clustered in groups of six or seven houses called *Baris* in dense tropical forest. Several closely located *Baris* comprise a *Samaj*. This type of settlement is less susceptible to severe cyclonic wind and storm surges.

Landless families tend to occupy coastal embankments illegally, as there are no public housing or welfare programs in Bangladesh for the landless. The respondents of our survey asserted that because of easy access to both land areas and the sea, they preferred to live on the embankment despite the well-known risks related to both illegal settlement and tropical cyclones (Figure 2).

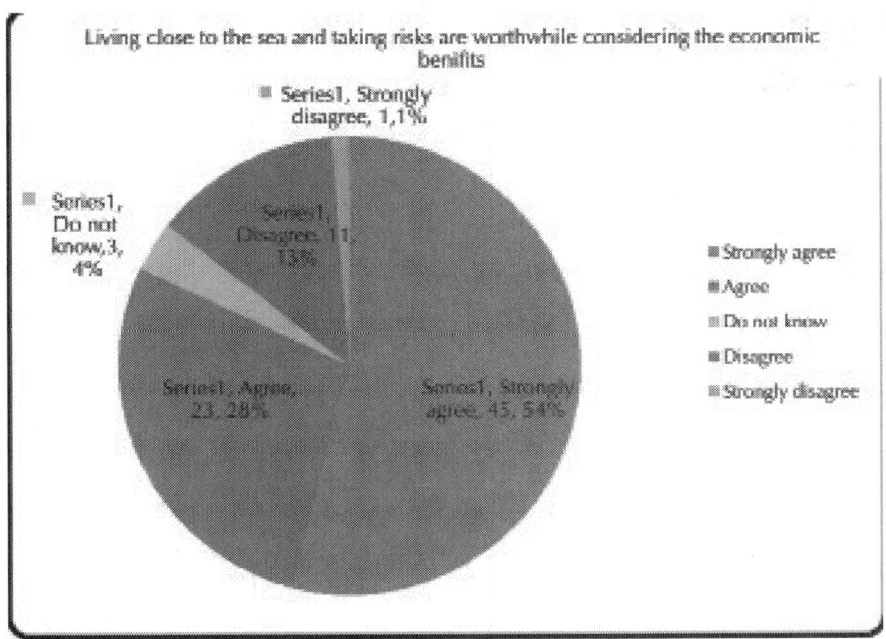

Figure 2: Consideration of villagers living close to the sea.

Cyclone Preparedness at the Local Level

Cyclone preparedness programs that have been implemented in recent years in the coastal zones of Bangladesh have involved both non-structural and structural measures. Appropriate cyclone preparedness training and enhancement of awareness by campaigns and public education have been major tools for building a well-prepared and cyclone-resilient community. Our field investigation revealed that only 31% of the local community members have received cyclone preparedness training during non-cyclone periods. Such training was chiefly provided by the non-governmental organizations (NGOs) that were locally active (30% of the local community members received training from them); national and local government initiatives in this regard were nominal (only 1% received training from them) (Figure 3).

Among the structural engineering measures, the construction of cyclone shelters and raised mounds (locally known as *killa*) to provide refuge during the onset of cyclones were the principal ones.

The expansion of coastal embankments and reforestation programs to protect settlements and properties from cyclone gusts and storm surges along the coast and estuary channels were among other significant structural measures. The majority (59%) of the community members took refuge in the designated cyclone shelters or in the masonry buildings of neighbors, friends, and relatives during the onset of Cyclone Sidr. About 41% stayed in their homes and opted not to go the cyclone shelters. Such behavior was attributed either to a strong belief that their lives were in the hands of *Allah* (i.e., strong presence of fatalism), a desire to save property from potential looting, the considerable distance of the shelters from their houses, or the unavailability of cyclone shelters in their locality.

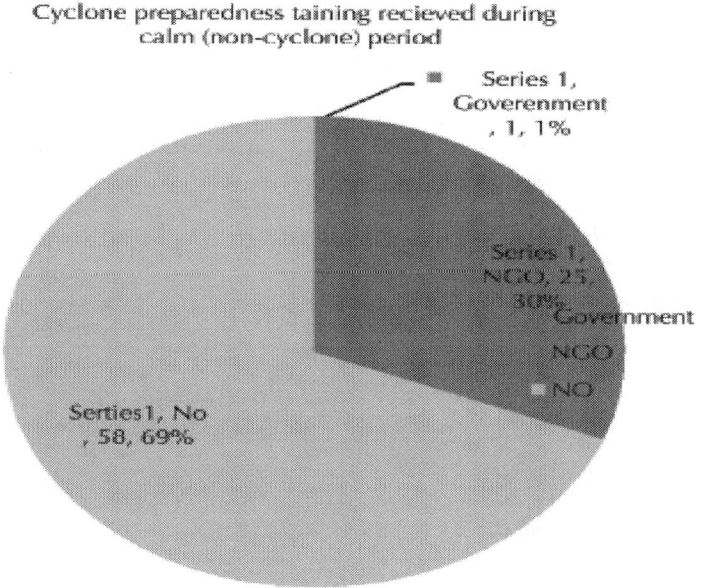

Figure 3: Training on cyclone preparedness.

The degree of variation in the number of people who took refuge per cyclone shelter among the localities was significant. Our field research calculated that there were 34 cyclone shelters for 631,138 people (according to the 2001 population census) in the Burguna Sadar *upazila*, under Burguna district, implying that each cyclone shelter would need to provide refuge to 18,563 people. In Kalapara *upazila*, under Patuakhali district, there were 113 cyclone shelters for 202,078 people, implying that each shelter would need to accommodate 1,788 people during the onset of a cyclone. Our exploration into why a considerable population did not use cyclone shelters during Cyclone Sidr provided a number of explanations. All cyclone shelters were being used as civic facilities (such as primary schools or community centers) during the non-cyclonic periods but lacked adequate drinking water supply infrastructure and toilets. A total of 87% of the cyclone shelter users identified low sanitation and inadequate drinking water facilities as major constraints to using these shelters. They also opined that low physical capacity (79%) and difficulty in maintaining privacy (47%) were other major problems in using the shelters (Table 2). The poor dispersion of cyclone shelters and the lack of road access were other major concerns about the cyclone shelters. The degree of satisfaction about the cyclone shelters was very low among the local community members (only 6% were satisfied). The examination of the use pattern of cyclone shelters at the local level revealed that although cyclone shelters saved many lives during Cyclone Sidr, many local people opted not to use them or they were simply unavailable in some localities. Clearly, these cyclone preparedness and mitigation measures are testimony of institutional partnerships at various levels of governance and public services.

Table 2: Problems in using cyclone shelters. Source: Field survey

Factors	Respondents (%) (n= 162)
Inadequate sanitation and drinking water facilities	87
Inadequate physical capacity	79
Hard to maintain privacy	47
Fragile structure	43
Unhygienic conditions	24
Absence of access road	14
Situated in a distant place	12
Do not know	5

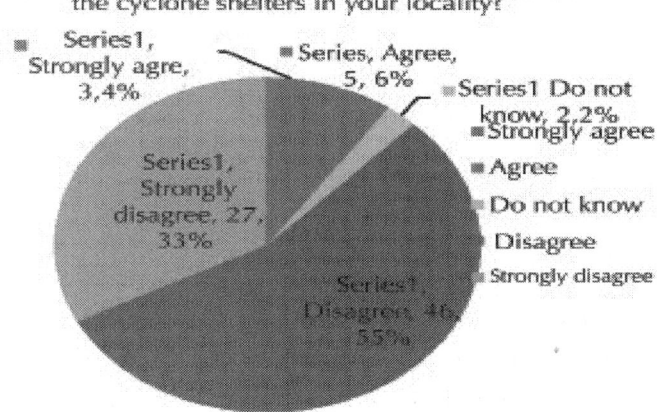

Figure 4: People's response on cyclone shelters and facilities.

Over the last four decades, the Water Development Board, in collaboration with other ministries, has constructed about 5,333 km of embankments in the coastal districts to support agriculture and protect the lives and property of coastal residents during cyclones and storm surges. Reducing vulnerability via these structural measures has been quite effective in coastal Bangladesh. There is clear evidence that these embankments along the coastal areas provided an effective buffer during the storm surge associated with Cyclone Sidr. Lives were saved, and damages and property losses were much lower where embankment structures had been properly maintained. Some embankments did not fail even when the storm surge overtopped them. More than 90% of our survey respondents observed that coastal embankments had saved their lives and property from Cyclone Sidr. They also registered that coastal plantations were another major living structural feature that reduced the velocity of the wind and the speed of tidal surges along with the cyclone and, in turn, saved their lives and properties.

CONCLUSIONS

The findings of our study reveal that, officially and legislatively, the government of Bangladesh in recent years has taken a comprehensive and integrated approach to disaster management. Both preparedness and response capacity have increased as a result. However, in the absence of stable and transparent institutions, this strong institutional partnership approach remains largely on paper. Individual stakeholders continue to make significant contributions, but synergy and multiplier effects are still missing. Our analysis shows that no, or only a very limited, culture of partnership in disaster management has yet been established. Divisive partisan politics and the lack of good governance prevent partnerships among stakeholders. Therefore, in the following section we propose a partnership framework that outlines new roles and responsibilities for major players. Implementation of the framework could lead to partnership in disaster management in Bangladesh. Government agencies, NGOs and policymakers need to understand the perspectives of local communities, the impacts of floods, and the levels of vulnerability to improve information, knowledge, resource support and services. For this, there is a need for more "action research" involving communities and scientists from different disciplines and greater awareness about the integration of floods, cyclones and other natural disasters and climatic events into the development process among key actors, particularly, government agencies.

Disaster management is a nationwide affair, involving each and every organization and citizen of the country. The government of any country cannot do it alone because of the resource constraints as well as the wide scope of the tasks involved. Therefore, a broad-based partnership involving all the stakeholders is a desirable and realistic approach to realizing the full potential at all stages of disaster management, namely, prevention, preparedness, response and recovery [31]. Experience that demonstrates the value of partnership in managing disasters is grounded in the mutual recognition of many different ideas and interests. The constituencies or interests associated with this partnership include the stakeholders, such as government ministries/agencies, National Parliament and the Parliamentary Standing Committee on Disaster Management, NGOs/CBOs, the private sector, the media, academia, donors and regional countries (Figure 5). The approach, involving multi-modal communication and interaction, proposes to integrate the activities of different stakeholders into a functional partnership framework. This outline describes how each stakeholder can be a loop in an integrated chain.

Ministry of Disaster Management and Relief (MoDMR)

This organization, directly supported by DMB and DRR, remains the pivot and is destined to channel and coordinate all communications and activities between and among all the partners in the loop. This coordinated process, particularly during non-crisis periods, is expected to result in the shift of focus from post-disaster to pre-disaster risk management. However, the MoDMR must ensure the transparency of its own and all other agencies involved in developing the partnership for disaster management. Public disclosure and documentation should be mandatory for all the stakeholders and must be published by this pivotal agency on a regular basis. The ultimate goal of the partnership is to enhance the investment and social capital for community empowerment against disaster risks.

From a partnership point of view, the following measures need to be initiated by the MoDMR:

- Adoption of a comprehensive national disaster management policy, with clear guidelines for an effective partnership of all stakeholders.

- Coordination of the functions of disaster management and climate change communities; this is likely to help integrate prevention with preparedness, response and recovery efforts, in both short-term and long-term perspectives; CDMP appears to be beginning in the right direction.

- Strengthening the capacity of the MoFDM and other disaster management-related agencies and committees at all levels, with particular emphasis on the district-level disaster management committees (DDMC). Each of the DDMC in the risk-prone areas can be equipped with a Geographic Information System (GIS) Cell as a planning tool for managing development and disaster reduction activities.

- Large-scale training of staff at all levels, including the stakeholders, particularly from the media, NGOs and private sector, in team and motivational works and in how to prevent disasters; for the purpose, a Disaster Management Training Cell can be established at the DMB.

- Activating the Disaster Management Committees (DMC) at all levels, including the national ones, through organizing meetings at regular intervals.

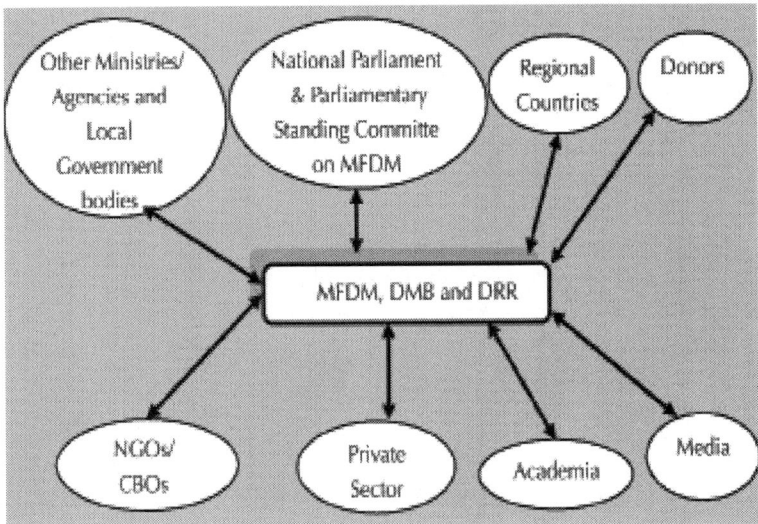

Figure 5: Proposed partnership framework for disaster management in Bangladesh.

- Strengthening the project monitoring and evaluation capacities at all levels, with the involvement of local stakeholders; the establishment of broad-based and inclusive monitoring and evaluation committees for projects will ensure transparency, accountability and therefore the delivery of intended results.

- Decentralization of not only responsibilities, but also decision-making power, to DDMCs, led by the local governments with sufficient financial resources and autonomy.

- Establishment of small teams at all levels of DMCs to better coordinate and integrate disaster management planning and activities from the national to local levels.

- Finally, developing a network among the GOs, NGOs, researchers, academics, journalists and other professionals in order to enlist their potential roles and contributions in mitigating disaster-related problems. The MoDMR can act as the coordinating agency for building up this network.

ACKNOWLEDGEMENTS

This research was funded by the Building Environmental Governance Capacity in Bangladesh (BEGCB) project under the financial support of Canadian International Development Agency (CIDA) and a grant from the Social Science and Humanities Research Council (SSHRC), Ottawa, Canada. The authors are thankful to the community members of Patuakhali and Barguna districts in Bangladesh for their participation and support to this research.

REFERENCES

1.   S Agrawala, T Ota, A. U Ahmed, J Smith, M. V Aalst, Development and Climate Change in Bangladesh: Focus on Coastal Flooding and the Sundarbans. Paris: Organization for Economic Co-operation and Development (OECD); 2003

2.   Government of BangladeshNational Adaptation Program of Action. Dhaka: Ministry of Environment and Forest; 2005

3.   C Loucks, S Barber-meyer, Hossain MAA, Barlow A, Chowdhury RM. Sea Level Rise and Tigers: Predicted Impacts to Bangladesh's Sundarbans Mangroves. Climatic Change 2010 98 291 298

4.   IPCC (Intergovernmental Panel on Climate Change)Climate Change 2001: The Scientific Basis. Contribution of Working Group-I to the IPCC Third Assessment Report, edited by Houghton et al. Cambridge University Press; 2001

5.   Asian Development Bank (ADB) and World BankBangladesh 2004 Post-Flood Recovery Programme: Damage and Needs Assessment. Dhaka; 2004

6.   J. R Chowdhury, R Rahman, Bangladesh Environment Outlook. Dhaka; 2001

7.   C. E Haque, Perspectives of Natural Disasters in East and South Asia, and the Pacific Island States: Socio-economic Correlates and Needs Assessment. Natural Hazards 2003 29 465 483

8.   CRED (Centre for Research on Epidemiology and Disasters)An International Disaster Database. Brussels: Université Catholique de Louvain; 2004http://www.emdat.be.

9. FFWC (Flood Forecasting and Warning Center)An Overview of Flood Forecasting and Warning Services in Bangladesh. A paper presented on 2nd April. Dhaka: Bangladesh Water Development Board; 2005

10. ISDR (International Strategy for Disaster Reduction)Living with Risk: A Global Review of Disaster Reduction Initiatives. Geneva; 2004http://www.unisdr.org/eng/about_isdr/bd-lwr-2004eng.htm.

11. CEGIS (Center for Environmental and Geographic Information Services)Bangladesh- Flood Affected Areas; 2007http://www.cegisbd.com/flood2007/index.htm.

12. OxfamSouth Asia Floods, 2007; 2007 http://www.oxfam.org/en/programs/emergencies/southasia_floods_07/update_070806.

13. GoB (Government of Bangladesh)Standing Orders on Disaster Management. Dhaka: Disaster Management Bureau; 1997

14. GoB (Government of Bangladesh)Poverty Reduction Strategy Paper (PRSP), December. Dhaka: Ministry of Finance and Planning; 2004

15. GoB (Government of Bangladesh)Disaster Management Act. Dhaka: Disaster Management Bureau; 1998

16. M. R Bhatt, T Reynolds, Community-Based Disaster Risk Reduction: Realizing the Primacy of Community. In: Haque CE, Etkin D. (eds.) Disaster Risk and Vulnerability: Mitigation through Mobilizing Communities and Partnerships. Montreal and Kingston, Canada: McGill and Queen's University Press; 2012 70 90

17. D. L Mallick, A Rahman, M Alam, Juel ASM, Ahmad AN, Alam SS. Bangladesh Floods in Bangladesh: A Shift from Disaster Management towards Disaster Preparedness. Institute of Development Studies Bulletin 2005 36 4 53 70

18. MoFDM (Ministry of Food and Disaster Management)Corporate Plan 2005-2009-Comprehensive Disaster Management: A Framework for Action. Dhaka: Government of Bangladesh; 2005

19. MoDMR (Ministry of Disaster Management and Relief) and UNDPDocuments on Comprehensive Disaster Management Project. Dhaka: Government of Bangladesh; 2004

20. T Cannon, Vulnerability Analysis and Natural Disasters. In: Varley A. (ed.) Disasters, Development and Environment. West Sussex, UK: Wiley; 1994

21. S Yodmani, Disaster Risk Management and Vulnerability Reduction: Protecting the Poor. Paper presented at The Asia and Pacific Forum on Poverty. Bangkok: Asian Disaster Preparedness Center; 2001p vi, 32.

22. LCG (Local Consultative Group)Notes on LCG Environment Meeting on Lessons Learned from Floods of 1998 and 2004, August 12, Dhaka; 2004

23. T Beck, Learning Lessons from Disaster Recovery: The Case of Bangladesh. Disaster Management Working paper Series 11Washington, DC: World Bank; 2005

24. C. E Haque, M. R Khan, M. S Uddin, S. R Chowdhury, Disaster Management and Public Policies in Bangladesh: Institutional Partnerships in Cyclone Hazards Mitigation and Response. In: Haque CE, Etkin D. (eds.) Disaster Risk and Vulnerability: Mitigation through Mobilizing Communities and Partnerships. Montreal and Kingston, Canada: McGill and Queen's University Press; 2012 154 182

25. BBCPlenty of agencies, but not enough aid. http://news.bbc. co.uk/2/hi/south_asia/stm, 19 November; 2007

26. BBCOne hundred days after SIDR-very little progress in rehabilitation and housing (an audio report in Bangla).http:// www.bbc.co.uk/bengali/indepth/story/2008/02/080222_ mbsidr_100days.shtml,February; 2008

27. A Maskrey, Reducing Global Disasters. In: Ingleton J. (ed.) Natural Disaster Management. Leicester, UK: Tudor Rose; 1999

28. C. E Haque, Atmospheric Hazards Preparedness in Bangladesh: A Study of Warning, Adjustments and Recovery from the April 1991 Cyclone. Natural Hazards 1997 16 181 202

29. B. K Paul, Why Relatively Fewer People Died? The Case of Bangladesh's Cyclone Sidr. Natural Hazards 2009 50 289 304

30. GoB (Government of Bangladesh)Cyclone Sidr in Bangladesh: Damage, Loss and Needs Assessment for Disaster Recovery and Reconstruction. Dhaka: Government of Bangladesh; 2008

31. E. L Quarantelli, Assessment of Development Potential and Capacity Based on Vulnerability. In: Integrated Approach to Disaster Management and Regional Development Planning with Peoples' Participation. UN Center for Regional Development; 1990

Conceptual Frameworks of Vulnerability Assessments for Natural Disasters Reduction

Roxana L. Ciurean[1], Dagmar Schröter[2], and Thomas Glade[1]

[1]Department of Geography and Regional Research, University of Vienna, Austria

[2]IIASA, Laxenburg, Austria

INTRODUCTION

The last few decades have demonstrated an increased concern for the occurrence of natural disasters and their consequences for leaders and organizations around the world. The EM-DAT International Disaster Database [1] statistics show that, in the last century, the mortality risk associated with major weather-related hazards has declined globally, but there has been a rapid increase in the exposure of economic assets to natural hazards.

Looking into more detail, UNISDR's Global Assessment Report 2011 (GAR11) [2] indicates that disasters in 2011 set a new record of $366 billion for economic losses, including $210 billion as a result of the Great East Japan Earthquake and the accompanying tsunami alone, and $40 billion as a result of the floods in Thailand. There were 29,782 deaths linked to 302 major disaster events including 19,846 deaths in the March earthquake/tsunami in Japan (figures presented by other disaster databases for 2011 summary e.g. NATCAT Service – MunichRE, are slightly different but in general agreement). Disaster databases, such as the ones referred to above, represent key resources for actors involved in policy and practice related with disaster risk reduction and response. However, considering their diversity and recognizing their different roles, one can identify at least one limitation in their use i.e. the inclusion criteria which inherently results in many hazard events not being registered. Compiling and analyzing an extensive natural disaster data set for the period 1993 – 2002, Alexander [3] showed that, for example, in the Philippines in 1996 there were 31 major floods, 29 earthquakes, 10 typhoons and 7 tornadoes. Due to population pressure, large areas of Luzon and other islands were denuded of their dense vegetation cover resulting in landslide prone slopes. Twelve major episodes of slope failure causing high damages to infrastructure and build up areas were registered in the archipelago during 1996. Although documentation of the Government expenditures to finance relief efforts for natural disasters during the 1996 – 2002 period is not completely contained in Figure 1 [4], one can observe that 1996 stands out as a particular year with high costs of rehabilitation.

Experience has shown that considering the frequency of disasters affecting the Philippines, its socio-economic context, and risk culture, the disaster management system tends to rely on a response approach. However, studies indicate that efforts are being made to engage more proactive approaches, involving mitigation and preparedness strategies [4]. In order to achieve this it is thus important to investigate not only the nature of the threat but also the underlying characteristics of the environment and society that makes them susceptible to damage and losses – in other words, the role of *vulnerability* in determining natural hazard risk levels.

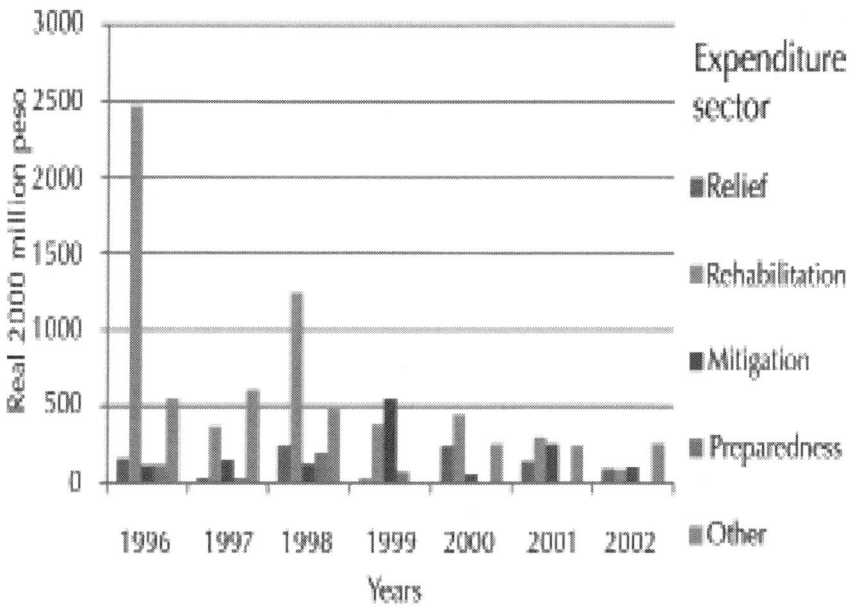

Figure 1: Philippines – annual expenditure under the National Calamity Fund (1996 – 2002) (Based on GDP at price market) [4].

BOX 1: Vulnerability – One Term Many Meanings

In everyday use of language, the term vulnerability refers to the inability to withstand the effects of a hostile environment. The definition of vulnerability for the purpose of scientific assessment depends on the purpose of the study – is it to get a differential picture of global change threats to human well-being in different world regions? Is it to inform particular stakeholders about adaptation options to a potential future development? Is it to show that likelihood of harm and cost of harm have changed for a specific element of interest within the human-environment system? In scientific assessment the term vulnerability can have many meanings, differentiated mostly by (a) the vulnerable entity studied, (b) the stakeholders of the study.

The design of scientific assessment (as opposed to scientific research) has to respond to the scientific needs of the particular stakeholder who might use it [5]. An integral part of vulnerability assessment therefore is the collaboration with its stakeholders [6], [7]. Thus, the specific definition and the method of vulnerability assessment is specific to each study and needs to be made transparent in the specific context. An example set of definitions on vulnerability used in natural hazards risk assessment and global change research is presented in section 2.2, Table 1.

The objective of this work is to discuss and illustrate different approaches used in vulnerability assessment for hydro-meteorological hazards (i.e. landslides and floods, incl. flash floods) taking into account two perspectives: hazard vulnerability and global change vulnerability, which are rooted in the technical and environmental as well as social disciplines. The study is based on a review of recent research findings in global change and natural hazards risk management. The overall aim is to identify current gaps that can guide the development of future perspectives for vulnerability analysis to hydro-meteorological hazards. Following the introduction (section 1), the second section starts with a definition of vulnerability within the context of risk management to natural hazards (sub-section 2.1). Subsequently, various conceptual models (sub-section 2.2) and vulnerability assessment methodologies (sub-section 2.3) are analyzed and compared based on their different disciplinary foci. In the third section, the importance of addressing uncertainty in vulnerability analysis is discussed and lastly general observations and concluding remarks are presented.

CONCEPTUAL FRAMEWORKS

Vulnerability and Risk Management to Natural Hazards

According to the UN International Strategy for Disaster Reduction (UNISDR) Report [8], there are two essential elements in the formulation of risk (Eq. 1): a potential event – hazard, and the degree of susceptibility of the elements exposed to that source – vulnerability.

$$RISK = HAZARD \times VULNERABILITY$$

(1)

In UNISDR terminology on Disaster Risk Reduction [9], «risk» is defined as the combination of the probability of an event and its negative consequences". A «hazard» is "a dangerous phenomenon, substance, human activity or condition that may cause loss of life, injury or other health impacts, property damage, loss of livelihoods and services, social and economic disruption, or environmental damage".

Within the risk management framework, vulnerability pertains to consequence analysis. It generally defines the potential for loss to the elements at risk caused by the occurrence of a hazard, and depends on multiple aspects arising from physical, social, economic, and environmental factors, which are interacting in space and time. Examples may include poor design and construction of buildings, inadequate protection of assets, lack of public information and awareness, limited official recognition of risks and preparedness measures, and disregard for wise environmental management.

BOX 2: Risk Management Frameworks are Generally Designed to Answer the Following Questions [10]:

What are the probable dangers and their magnitude? (Danger Identification)

How often do the dangers of a given magnitude occur? (Hazard Assessment)

What are the elements at risk? (Elements at Risk Identification)

What is the possible damage to the elements at risk? (Vulnerability Assessment)

What is the probability of damage? (Risk Estimation)

What is the significance of the estimated risk? (Risk Evaluation)

What should be done? (Risk Management)

Vulnerability Models

There are multiple definitions, concepts and methods to systematize vulnerability denoting the plurality of views and meanings attached to this term. Birkmann [11] noted that 'we are still dealing with a paradox: we aim to measure vulnerability, yet we cannot define it precisely'. However, there are generally two perspectives in which vulnerability can be viewed and which are closely linked with the evolution of the concept [12]: (1) the amount of damage caused to a system by a particular hazard (technical or engineering sciences oriented perspective – dominating the disaster risk perception in the 1970s), and (2) a state that exists within a system before it encounters a hazard (social sciences oriented perspective – an alternative paradigm which uses vulnerability as a starting point for risk reduction since the 1980s). The former emphasizes 'assessments of hazards and their impacts, in which the role of human systems in mediating the outcomes of hazard events is downplayed or neglected'. The latter puts the human system on the central stage and focuses on determining the coping capacity of the society, the ability to resist, respond and recover from the impact of a natural hazard [13]. While the technical sciences perspective of vulnerability focuses primarily on physical aspects [14], the social sciences perspective takes into account various factors and parameters that influence vulnerability, such as physical, economic, social, environmental, and institutional characteristics [8]. Other approaches emphasize the need to account for additional global factors, such as globalization and climate change. Thus, the broader vulnerability assessment is in scope, the more interdisciplinary it becomes.

Table 1: General definitions of vulnerability used in risk assessment due to natural hazards and climate change

Working definitions(s): Vulnerability is...	Source
The degree of loss to a given element at risk or a set of elements at risk resulting from the occurrence of a natural phenomenon of a given magnitude and expressed on a scale from 0 (no damage) to 1 (total damage)	[14]

The conditions determined by physical, social, economic, and environmental factors or processes, which increase the susceptibility of a community to the impact of hazards	[8]
The characteristics of a person or group in terms of their capacity to anticipate, cope with, resist and recover from impacts of a hazard	[13]
The intrinsic and dynamic feature of an element at risk that determines the expected damage/harm resulting from a given hazardous event and is often even affected by the harmful event itself. Vulnerability changes continuously over time and is driven by physical, social, economic and environmental factors	[11]
The degree to which geophysical, biological and socio-economic systems are susceptible to, and unable to cope with, adverse impacts of climate change	[15], [16]

The different definitions of vulnerability can also be viewed from a functional and subject/object-oriented perspective i.e. considering the end-user of the scientific assessment results (e.g. technical boards, administration officers, representatives from the civil protection, international organizations, etc.) and the vulnerable entity (e.g. critical infrastructure, elderly population, communication networks, mountain ecosystems, etc.).

Vogel and O'Brien [17] emphasize that vulnerability is: *(a) multi-dimensional and differential* (varies for different dimensions of a single element or group of elements and from a physical context to another); *(b) scale dependent* (with regard to the unit of analysis e.g. individual, local, regional, national etc.) *and (c) dynamic* (the characteristics that influence vulnerability are continuously changing in time and space). With regards to the first characteristic, there are generally five components (or dimensions) that need to be investigated in vulnerability assessment: (1) the physical/functional dimension (relates to the predisposition of a structure, infrastructure or service to be damaged due to the occurrence of a harmful event associated with a specific hazard); (2) the economic dimension (relates to the economic stability of a region endangered by a a loss of production, decrease of income or consumption of goods due to the occurrence of a hazard); (3) the social dimension (relates with the presence of human beings, individuals or communities, and their capacities to cope with, resist and recover from impacts of hazards); (4) the environmental dimension (refers to the interrelation between different ecosystems and their ability to cope with and recover from impacts of hazards and to tolerate stressors over time and space); (5) the political/institutional dimension (refers to those political or institutional actions e.g. livelihood diversification, risk mitigation strategies, regulation control, etc., or characteristics that determine differential coping capacities and exposure to hazards and associated impacts).

During the last decades, various schools of thinking proposed different conceptual models with the final aim of developing methods for measuring vulnerability. The following sub-sections give a short overview of some of the conceptual models presented in [11], such as the double structure of vulnerability, vulnerability within the context of hazard and risk, vulnerability in the context of global environmental change community, the Presure and Release Model and a holistic approach to risk and vulnerability assessment. Other models not discussed herein are: The Sustainable Livelihood Framework, the UNISDR framework for disaster risk reduction, the 'onion framework', and the 'BBC conceptual framework', the last two developed by UNU-EHS (UN University, Institute for Environment and Human Security).

The Double Structure of Vulnerability

According to Bohle [18] vulnerability can be seen as having an external and internal side (Figure 2). The *external* side is related to the exposure to risks and shocks and is influenced by Political Economy Approaches (e.g. social inequities, disproportionate division of assets), Human Ecology Perspectives (population dynamics and environmental management capacities) and the Entitlement Theory (relates vulnerability to the incapacity of people to obtain or manage assets via legitimate economic means). The *internal* side is called coping and relates to the capacity to anticipate, cope with, resist and recover from the impact of a hazard and is influenced by the Crisis and Conflict Theory (control of assets and resources, capacities to manage crisis situations and resolve conflicts), Action Theory Approaches (how people act and react freely as a result of social, economic or governmental constrains) and Model of Access to Assets (mitigation of vulnerability through access to assets). The conceptual framework of the double structure indicates that vulnerability cannot adequately be considered without taking into account coping [1] - and response capacity [2] - .

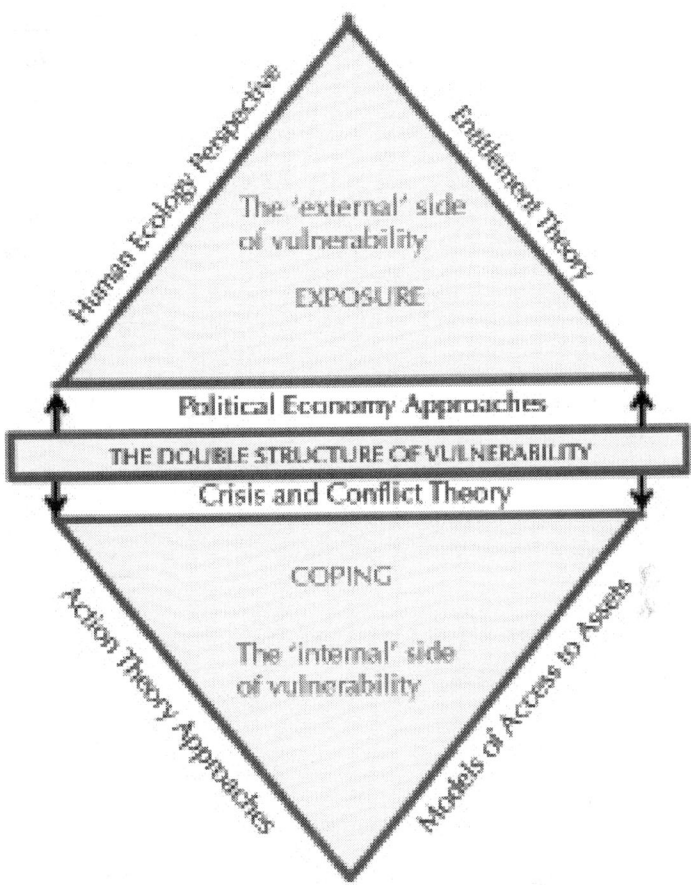

Figure 2: Bohle's conceptual framework for vulnerability analysis [18] in [11].

Vulnerability within the Framework of Hazard and Risk

The disaster risk community defines vulnerability as a component within the context of hazard and risk. This school usually views vulnerability, coping capacity and exposure as separate features. One example within this approach is Davidson's [19] conceptual framework, adopted in [20] and illustrated in Figure 3. This framework views risk as the sum of hazard, exposure [3] - , vulnerability and capacity measures. Hazard is characterized by probability and severity, exposure is characterized by structure, population and economy, while vulnerability has a physical, social, economic and environmental dimension. Capacity and measures are related with physical planning, management as well as social – and economic capacity.

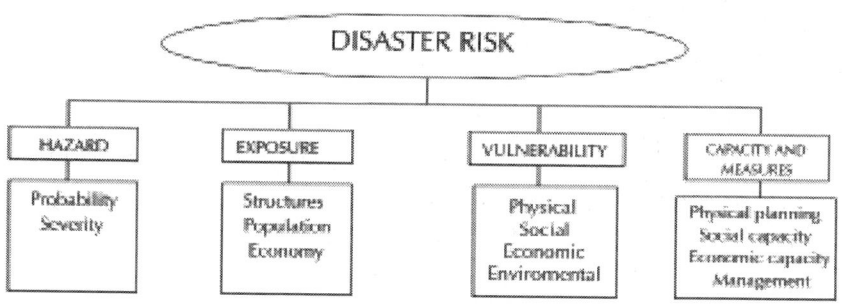

Figure 3: Conceptual framework to identify risk [20] in [11].

Vulnerability in the Global Environmental Change Community

Turner [21] developed a conceptual framework considered representative for the global environmental change community primarily due to its focus on the coupled human-environment systems. Their definition of vulnerability encompasses exposure, sensitivity and resilience. Exposure contains a set of components (i.e. threatened elements: individuals, households, states, ecosystem, etc.) subjected to damage and characteristics of the threat (frequency, magnitude, duration). The sensitivity is determined by the human (social capital and endowments) and environmental (natural capital or biophysical endowments) conditions of the system which influence its resilience [4] - . The last component is enhanced through adjustments and adaptation.

A system's vulnerability to hazards consists of (Figure 4) (i) linkages to the broader human and biophysical (environmental) conditions and processes operating on the coupled system in question; (ii) perturbations and stressors/stresses [5] - that emerge from this conditions and processes; and (iii) the coupled human – environment system of concern in which vulnerability resides, including exposure and responses (i.e. coping, impacts, adjustments, and adaptation) [21].

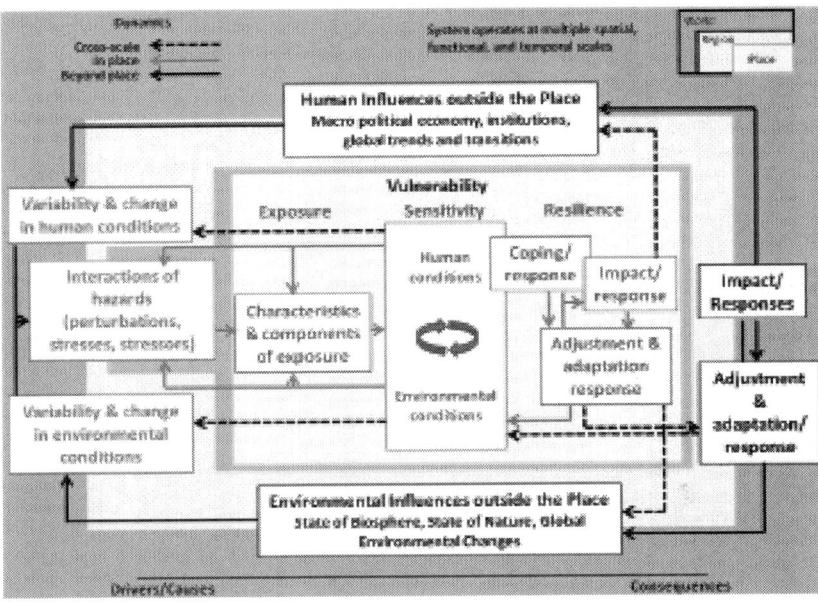

Figure 4: Vulnerability conceptual framework [21] in [11].

The Pressure and Release Model (Par Model)

The model operates at different spatial (place, region, world), functional and temporal scales and takes into account the interaction of the multiple perturbations and stressor/stresses [22]. Hazards are regarded as being influenced from inside and outside of the analyzed system; however, due to their character they are commonly considered site-specific. Thus, given their complexity, hazards are located within and beyond the place of assessment. The Pressure and Release model (PAR model) is based on the commonly used equation which defines risk as a function of the hazard and vulnerability (Eq. 1). It emphasizes the underlying driving forces of vulnerability and the conditions existent in a system that contribute to disaster situations when a hazard occurs. Vulnerability is associated with these conditions at three progressive levels: (1) *Root causes*, which can be, for example, limited access to power, structures or resources; or related with political ideologies or economic systems; (2) *dynamic pressures*represented, for example, by demographic or social changes in time and space (e.g. rapid population decrease, rapid urbanization, lack of local institutions, appropriate skills or training); and (3) *unsafe conditions* posed by the physical environment (e.g. unprotected buildings and infrastructure, dangerous slopes) or socio-economic context (e.g. lack of local institutions, prevalence of endemic diseases). In Birkmann's opinion [11], this conceptual framework is an important approach which goes beyond identification of vulnerability towards addressing its root causes and driving forces embedded in the human-environment system.

A Holistic Approach to Risk and Vulnerability

In this approach vulnerability is conditions by three categories of factors [23]:

- Physical exposure and susceptibility – regarded as hazard dependent
- Fragility of the socio-economic system – non hazard dependent
- Lack of resilience to cope and recover – non hazard dependent

The authors emphasize the importance of measuring vulnerability from a comprehensive and multidisciplinary perspective. The model (Figure 5) takes into account the consequences of direct physical impacts (exposure and susceptibility) as well as indirect consequences (socio-economic fragility and lack of resilience) of potential hazardous event. Within each category, the vulnerability factors are described with sets of indicators or indices. The model includes a control system which alters indirectly the level of risk through corrective and prospective interventions (risk identification, risk reduction, disaster management).

Physical Vulnerability Assessment

If in social vulnerability assessment the focus is on determining the indicators of societies' coping capacities to any natural hazard and identifying the vulnerable groups or individuals based on these indicators, in physical (or technical) vulnerability assessment the role of hazard and their impacts is emphasized, while the human systems in mediating the outcomes is minimized. In the technical/engineering literature for natural hazards, physical vulnerability is generally defined on a scale ranging from 0 (no loss/damage) to 1 (total loss/damage), representing the degree of loss/potential damage of the element at risk (see Table 1). The evaluation of vulnerability and the combination of the hazard and the vulnerability to obtain the risk differs between natural phenomena. However, the majority of models see vulnerability as being dependent both on the acting agent (physical impact of a hazard event) and the exposed element (structural or physical characteristics of the vulnerable object). The most common expressions of physical vulnerability for different types of hazards (landslides, floods, earthquakes) are: vulnerability curves (stage-damage functions), fragility curves, damage matrices and vulnerability indicators [35]. In recent decades, research on flood vulnerability assessment has advanced substantially (especially with the aid of computational techniques) and different modeling approaches ranging from post-event damage observations to laboratory-based experiments and physical modeling have been developed. One major applications of flood vulnerability analysis is the development of guidelines for reducing structural vulnerability for different types of properties. Likewise, the results of these studies are used in spatial development strategies (spatial planning) and for identification of the elements or areas where damages would be expected in case of flood occurrence. There are two main approaches of flood vulnerability assessment: one (1) focuses on the economic damage and is essentially a quantification of the expected or actual damages to a structure expressed in monetary terms or through an evaluation of the percentage of the expected loss; (2) the other, deals with the physical vulnerability of individual structures and on the estimation of the likelihood of occurrence of physical damages or collapse of a single element (e.g. a building). Within the last category, two general methods can be identified:

Empirical methods are based on the analysis of observed consequences (collection of actual flood damage information after the event) through the use of interviews, questionnaires and field mapping. The main advantage of these methods is the use of real data. However, the results are very much dependent on the respondents' risk perception for the first two - and data availability (especially for deriving stage-damage curves) for the last collection method.

In analytical methods (i) different flood parameters (duration, velocity, impact pressure, etc.) are directly controlled during laboratory experiments and their effects on the structures are quantified; (ii) numerical models and computer simulation techniques are used to estimate the reliability of a structure and/or calculate its probability of failure (usually hydrologic and hydraulic modeling of the floodplain is a pre-requisite) [36]. This type of approaches are resource demanding (time and money) but allow for a better understanding of the relation between flood intensity and degree of damage for an exposed structure with definite characteristics. Moreover, due to data/resources requirement, they can only be used for assessment of individual structures.

The key parameters used in order to quantify physical vulnerability to floods are related with the forces (buoyancy, hydrostatic pressure and dynamic pressure) that flooding is likely to exert on a structure (e.g. building, bridge, dam, etc.). Directly linked with these forces are the characteristics of the damaging agent (water) which are reflected in a number of actions on the exposed structure: hydrostatic, hydrodynamic, erosion, buoyancy, etc. ([37] in [38]).

The most used approach for assessing and modeling direct flood damages is the stage-damage functions which relates the relative or absolute damage for a certain class of objects to the inundation depth (Figure 7). One limitation in their use is the assessment of the degree of damage based solely on one characteristic of the exposed element/group of elements (e.g. building type). Likewise, the flood damage influencing parameter e.g. inundation depth, may not be the only hazard indicator that contributes to the quantity of losses [39]. In [40] the importance of further influencing factors like 'duration of inundation, sediment concentration, availability and information content of flood warning and the quality of external response in a flood situation' are emphasized. For static floods (slow moving water) the depth is considered to be sufficient for the analysis, but for dynamic floods, velocity is regarded as more important.

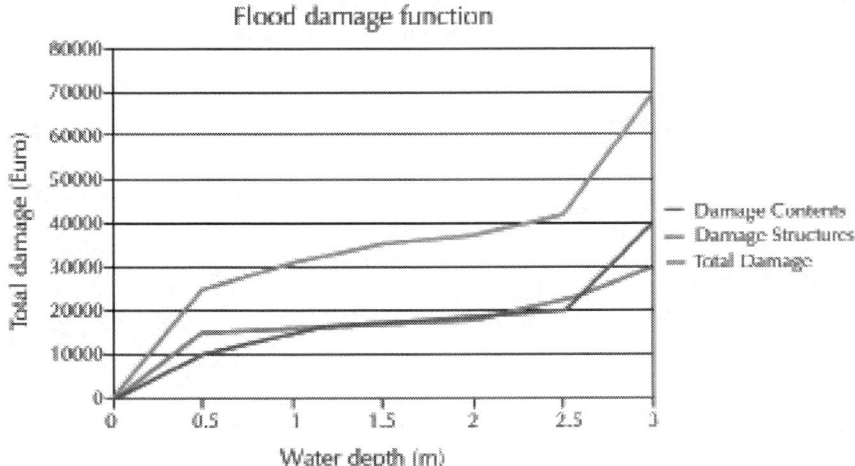

Figure 7: Example of flood damage curves showing damage to structures, contents and total damage as a function of inundation depths [41].

In HAZUS-MH Flood Model [42] the latter parameter is directly considered. A velocity-depth function is included indicating if building collapse has to be assumed. A threshold for collapse corresponding to 100% damage is set, while below this threshold the damage is estimated based on the inundation level only. The model also takes into account the effect of warning which is assessed based on a 'day-curve'. If a public response rate of 100% is assumed, a maximum of 35% of damage reduction can be achieved depending on the time of warning [26]. The flood hazard module addresses both riverine and coastal floods; flash-floods are not included in the model's capability.

The Swiss risk concept from the Nationale Platform Naturgefahren (PLANAT) defines three intensity classes for flood vulnerability analysis, based on flood depth and velocity which are used in spatial planning regulations (Table 2).

Table 2: Intensity classes based on flood depth and velocity from PLANAT in [26]

Intensity class	Criteria	Description
Low	h < 0.5 m or v x h < 0.5 m2/s	Persons are barely at risk and only low damages at buildings or disruption have to be expected
Middle	2 m > h > 0.5 m or 2 m2/s > v x h > 0.5 m2/s	Persons outside of buildings are at risk and damage to buildings can occur while persons in buildings are quite safe and sudden destruction of buildings is improbable
High	h > 2 m or v x h > 2 m2/s	Persons inside and outside of buildings are at risk and the destruction of buildings is possible or events with lower intensity occur but with higher frequency and persons outside of buildings are at risk

Damages caused by landslides to population, environment and built-up areas are significantly less than for other natural hazards due to the inherent characteristic of the process. However, the extent of these losses is frequently underestimated especially when landslides are associated with the occurrence of floods or earthquakes (their consequences tend to be aggregated). Generally, vulnerability to landslides depends on a variety of factors like: runout distance; volume and velocity of sliding; pressure caused by the movement; height of deposition; elements at risk (e.g. different structures), their nature and their proximity to the slide; elements at risk (e.g. persons), their proximity to the slide, the nature of the building/roads they are in [43].

Research in the field of landslide hazard and risk ([24], [44], [45], [46]) has demonstrated that in contrast to other natural processes (flooding, earthquakes) landslide vulnerability is more difficult to assess due to a number of reason, such as:

- The complexity and the wide range of variety of landslide processes (landslides are determined by different predisposing and triggering factors which results in various mechanisms of failure and mobility, size, shape, etc.)

- The lack of systematic methods for expressing landslide intensity - there is no general indicator of landslide intensity (e.g. for rock falls, impact pressure or volume can be used whereas for debris flow deposit height is common; other indicators such as flow velocity are rarely considered) and in practice data scarcity reduces their number significantly

- The quantitative heterogeneity of vulnerability of different elements at risk for qualitatively similar landslide mechanisms due to their intrinsic characteristics (here, human life constitutes a special case)

- The variability in spatial and temporal vulnerability

- The lack of historical damage databases – usually only events which cause extensive damage are recorded and data about the type and extent of damage is often missing

- Non-physical factors influence the vulnerability of people (e.g. early warning, hazard and risk perception, etc.)

Landslide vulnerability assessment approaches range significantly due to various foci and objectives addressed. Some consider vulnerability within the landslide risk management framework, others evaluate exclusively physical vulnerability. Three general types of methodologies can be identified (without excluding the possibility of other classification schemes):

Qualitative Methods ([47], [48], [35]) - given a particular landslide type and the characteristics of the elements at risk, the appropriate vulnerability factor is assessed by expert judgment, field mapping or based on historical records. These methods are flexible (e.g. indicator based methods) valuable and easy to use/understand by decision makers. However, a major limitation of this approach is that most of the data have to be assumed and there is no direct (quantified) relation between hazard intensities and degree of damage.

As an example, in [47] an empirical GIS-based geomorphological approach for landslide and risk analysis was proposed, using stereoscopic aerial photographs and field mapping in order to represent the changes in distribution and shape of landslides and assess their expected frequency of occurrence and intensity. The damages were classified in three classes using a qualitative relationship between landslide intensity/type and their consequences: *superficial* (aesthetic, minor) damage where the functionality of the elements at risk is not compromised and damage can be repaired, rapidly and at low costs; *functional* (medium) damage, where the functionality of the structures is compromised, and the damage takes time and large resources to be fixed; *structural* (total) damage, where buildings or transportation routes are severely or completely damaged, and require extensive (and costly) work to be fixed (demolition and reconstruction may be required).

Semi-Quantitative Methods are reducing the level of generalization in comparison with qualitative methods. They are flexible and can, to a certain degree, reduce subjectivity, compared with the methods mentioned above. Within this category, damage matrices, for example, are composed by classified intensities and stepwise damage levels. In [49] damage matrices were suggested based on damaging factors and the resistance of the elements at risk to the impact of landslides. Figure 8 shows a correlation, in terms of vulnerability, between exposed elements and the characteristics of the hazard. The applicability of this method, requires statistical analysis of detailed records on landslides and their consequences [50]. This proves to be a challenge in data scarce environments.

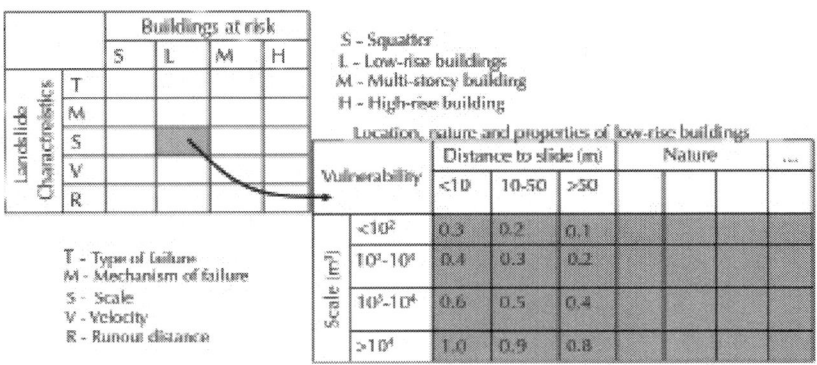

Figure 8: Structural vulnerability matrix [49].

Quantitative Methods ([51], [52], [53], [54]) are mostly applied at local scale (often, for individual structures) due to complexity of procedures involved and detailed data requirements. Quantitative methods are usually employed by engineers or actors involved in technical decision making, as they allow for a more explicit objective output. The results can be directly integrated in a Quantitative Risk Assessment (QRA) also taking into account the uncertainty in vulnerability analysis. The procedures involved can rely on i) expert judgment (heuristic), ii) damage records (empirical) or iii) statistical analysis (probabilistic).

One example of quantitative expert judgment used to evaluate physical vulnerability of roads to debris flows was used in [55]. 147 respondents from 17 countries were asked to use their expert knowledge to assess the probability of a certain damage state being exceeded given that a volume of debris impacts a road (Table 3).

Table 3: Damage state definition [55]

Description of probabilities		
Descriptor	Description	Values for analysis
Highly improbable	Damage state almost certainly exceeded, but cannot be ruled out	0.000001
Improbable(remote)	Damage state only exceeded in exceptional circumstances	0.00001
Very unlikely	Damage state will only be exceeded in very unusual circumstances	0.001
Unlikely	Damage state may be exceeded, but would not be expected to occur under normal circumstances	0.001
Likely	Damage state expected to be exceeded	0.01
Very likely	Damage state almost certainly exceeded	0.1

Based on the questionnaire results, fragility curves were produced which relate the flow volume to damage probabilities (Figures 9). It should be noted that in this study probabilites were derived based on the respondents experience only (qualitative data) with no statistical processing of damage observations or analytical/numerical modeling. The results were compared to known events in Scotland (UK) and the

Republic of Korea. The major limitation of this method is the high degree of subjectivity, however it advances expert knowledge which might be in some cases the only/most appropriate source of information about damages caused by the impact of landslides.

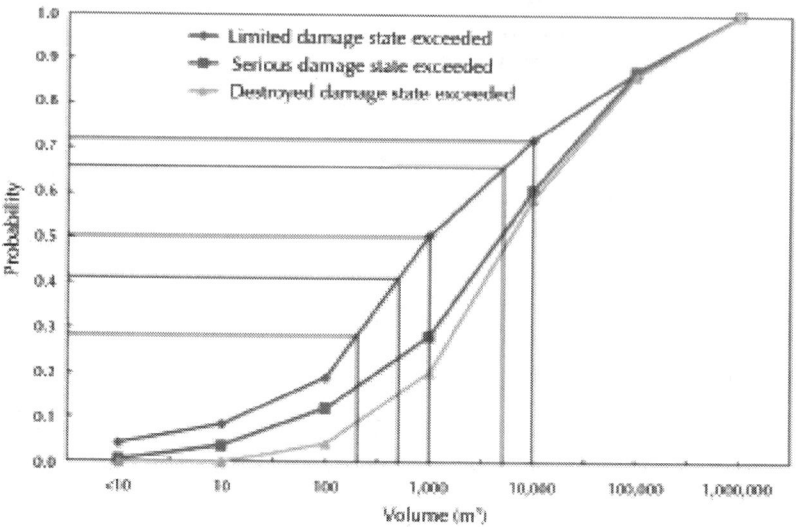

Figure 9: Fragility curves 'forced' to unity and manually extrapolated to the next order of magnitude for volume (local roads). The vertical lines are added at 200, 500, 1000, 5000 and 10000 m³ (illustration only for 'limited damage' curves) [55].

In reference [53], the author performed a study of a well-documented debris flow event which occurred in the Austrian Alps (August, 1997) and derived vulnerability curves for buildings located on the fan of the torrent based on the intensity of the phenomenon and the damage ratio. The intensity was approximated by deposit height and the susceptibility of the element at risk (i.e. buildings) by material of construction (brick, masonry, and concrete). Figure 10 shows the curve produced together with other existing curves for comparison. The application of this vulnerability function is limited to process intensities expressed as deposit height ≤ 2.5 – 3 m which means that the curve is not relevant for intensities which exceed this value. Nevertheless, the authors argue that such high process intensities generally result in a total loss of the building since the reparation costs will exceed the expenditure necessary for a new construction [53].

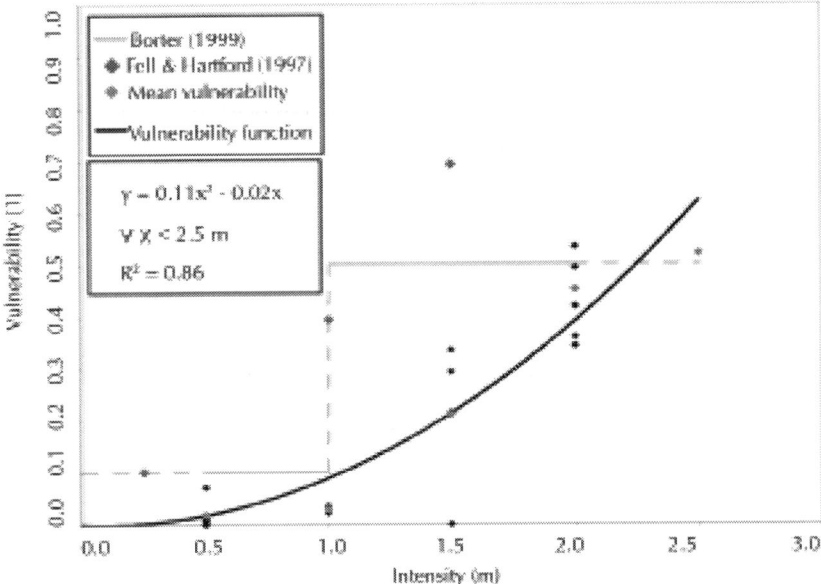

Figure 10: Relationship between debris flow intensity and vulnerability is expressed by a second order polynomial function for flow height > 2.5 m. Results from the study are indicated by black dots, the corresponding mean vulnerability is indicated by red dots [53].

In another study [51], a scenario-based method derived from a probabilistic approach to regional vulnerability assessment [56] was used. The authors defined vulnerability as a function of landslide intensity and the susceptibility of vulnerable elements (see Eq. 2).

$$V = I \bullet S$$

(2)

Susceptibility is defined as 'the lack of inherent capacity of the elements in the spatial extension under investigation to preserve their physical integrity and functionality in the course of the physical interaction with a generic sliding mass' and is independent of the characteristics of the landslide [51]. The susceptibility model is able to accommodate any factor dictated by the analyzed category of elements at risk. In this study, the susceptibility factors taken into account are: (a) resistance and state of maintenance for structures, and (b) persons in open space and vehicles, population density, income, age, and persons in structures, for individuals. For landslide intensity, a composite parameter is derived based on the kinetic – (related with the damage caused by the impact energy of the sliding mass) and kinematic (accounts for the effects of size-linked features of a reference landslide) characteristics of the interaction between the sliding mass and the reference area proposed. Models for quantification of susceptibility (Eq. 2) and intensity (Eq. 3) are illustrated below:

$$S = 1 - \prod_{i=1}^{ns} (1 - \vartheta i)$$

(3)

Where, ϑ_i is the ith on ns susceptibility factor (each defined in the range) contributing to the category susceptibility and,

$$I = ks \bullet (rK \bullet IK + rM \bullet IM)$$

(4)

Where, k_s is the spatial impact ratio (equal to the ratio between the area pertaining to the category that is affected by the landslide and the total area pertaining to the category); $_rK$ and IK are kinetic factors and $_rM$ and IM are kinematic factors. The proposed methodology provided a framework for the quantification of uncertainties in vulnerability assessment.

UNCERTAINTY IN VULNERABILITY ANALYSIS

In natural hazards risk management, decisions regarding the risk associated with a particular hazard are essentially enacted based on limited information and resources. In order to improve this process, experts started to investigate the effects of uncertainty on risk (and its determinants) qualitatively or quantitatively and communicate their results to decision-makers. This one-way approach toward finding solutions for advancing decision making proves out to be insufficient in contrast to the complexity of the problems at hand, especially when dealing with inherent uncertainties or unforeseen changes in the human-environmental system. Nevertheless, effort are being made to reduce the effects of uncertainty on vulnerability (and consequently, risk), particularly related with the data and models used. For example, representing hazard damage potential by only one parameter (e.g. for floods – depth of inundation) can result in overestimations of vulnerability and subsequently in un-economic investments in mitigation countermeasures. One possibility to overcome this problem would be to reduce the uncertainty in the input data by using data-mining approaches (e.g. tree-structured models) for the selection of the most important damage-influencing parameters [39]. Other examples would be the use of scenario analysis for seismic vulnerability and its probable damages in order to develop a hierarchy of effective factors in earthquake vulnerability [57] or testing the performance of different structures (reliability analysis) subjected to the impact of landslides with various intensities through the use of traditional methods like Monte Carlo Simulation (MCS), First Order Second Moment (FOSM), First Order - /Second Order Reliability Method (FORM/SORM). However, the selection of the most appropriate uncertainty modeling approach depends on the level of complexity required by the scope of analysis or the use of the final results.

Generally, uncertainties in decision and risk analysis can be divided into two categories [10]: those that stem from 'real' variability in known (or observable) processes or phenomena (e.g. height or the ethnicity of an arbitrary individual in a specified population or the distribution of velocities in a sliding mass, etc.) and those which reside from our limited knowledge about fundamental phenomena (e.g. the nature of some earthquake mechanism, the effect of water level fluctuation on clay slope stability, etc.). The former is known as aleatory (inherent or stochastic) uncertainty and cannot be reduced. The latter, epistemic uncertainty, includes measurement uncertainty, statistical uncertainty (due to limited information), and model uncertainty, which can be reduced, for example, by increasing the probing samples or by improving the measurement methods or modeling algorithms. Other types of classification systems, together with a review of methods and simulation techniques for uncertainty treatment are critically discussed and illustrated in a work performed by the Norwegian Geotechnical Institute (NGI), in [34]. Uncertainty can be addressed from (1) an integrative perspective, where vulnerability is registered by exposure to hazards but also resides in the resilience of the system experiencing the hazard [58] (bottom-up oriented vulnerability assessment). In this context, uncertainty is associated with future changes (in frequency and magnitude of hazards but also in climatic, environmental and socio-economic patterns) characterized by unknowable risks to which communities must learn to adapt. This approach is centered on the human systems' coping capacity and promotes vulnerability reduction through enhancing resilience to future change. Conversely, (2) a direct approach towards reduction of (epistemic) uncertainty is developed within the technical field (assimilated to deductive, top-down vulnerability assessments), where uncertainty models are defined for each component of vulnerability and the sources of uncertainty categorized [45]. Figure 11 shows how these two approaches of dealing with uncertainty can inform climate adaptation policy: one is (epistemic) uncertainty 'reducer' while the other is uncertainty 'accepting' (due to issues like, for example, timescale and planning horizons, the unit of analysis being considered and the development status of the region or country) [59].

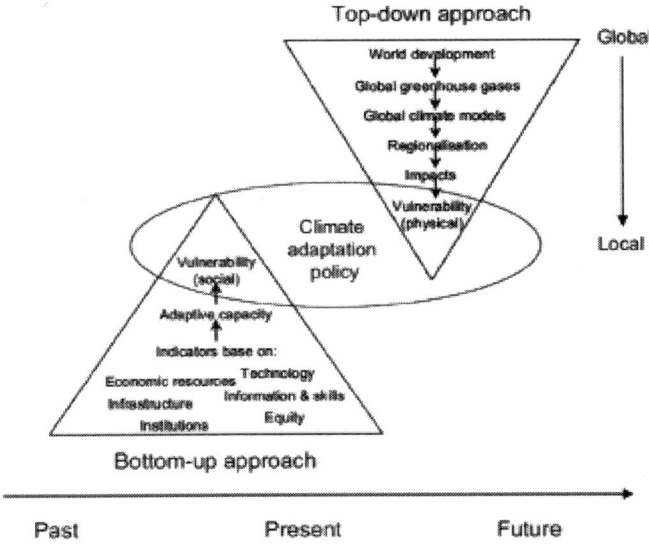

Figure 11: "Top-down" and "bottom-up" approaches used to inform adaptation to climate change [59].

Table 4 illustrates an example of uncertainty sources in physical vulnerability analysis of buildings. It is obvious that these will vary with the methodology used and the quality and quantity of data available.

Table 4: Sources of uncertainty in physical vulnerability to landslides (e.g. for buildings)

Type	Source
Epistemic	Intensity assessment (using proxies e.g. depth of material, velocity, volume, impact pressure, etc.) Characterization of elements at risk (e.g. structural-morphological characteristics, state of maintenance, strategic relevance, etc.) Estimations of buildings' value and damage costsVulnerability model (selection of parameters, mathematical model, calculation limitations) Expert judgement
Aleatory	Spatial variability of parameters* (e.g. landslide intensities, population density, etc.)

[i] - *also related with the scale of investigation

Within the general risk assessment framework, uncertainty propagates not only from one component of risk to another but also within the process stages of vulnerability analysis. This is schematically described in a classification system for vulnerability estimation proposed in [34] and represented inFigure 12.

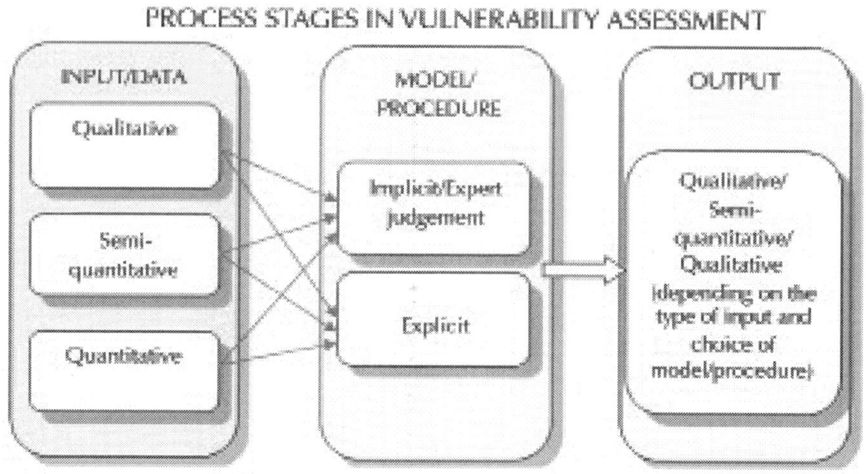

PROCESS STAGES IN VULNERABILITY ASSESSMENT

Figure 12: Classification system for vulnerability estimation. Uncertainty is associated with each process stage [34].

According to the authors, uncertainty associated with the input data (depending on the type, quantity and quality), propagates through the model, which also contains a degree of uncertainty due to, for example, expert judgment, mathematical model or basic assumptions. The uncertainty in the output depends on the two previous process stages as well as the uncertainty related with the interpretation of the results.

CONCLUSIONS

The most important goal in developing tools for measuring vulnerability is their use in natural hazards risk reduction strategies, thus applying them in decision making processes. In this context, it is necessary to know what is the objective of the assessment, what is the target group of any particular approach, who is using the results and what is their understanding of the outcome. The methods of vulnerability assessment presented herein are mere exemplification of the complexity and wide range of approaches that can be applied in natural hazards disaster risk management. However, based on these a number of observations may be formulated.

Vulnerability defined considering physical exposure or social-economical determinants only cannot encompass the complexity of effects caused by the impact of a natural hazard on an element or group of elements at risk (especially for systems like urban developments, communities, etc.). In an editorial for vulnerability to natural hazards [60] addressed the question of integration between natural and social scientific approaches based on a number of research studies. Their findings show that, studies that are dedicated to different components of vulnerability (e.g. frequency and magnitude of a hazard, elements at risk, exposure, coping and adaptation capacities, etc.) and therefore use different methodological approaches, are relatively similar in scope. Hence it is important to clearly describe and define which components of risk and/or vulnerability assessment are considered in each individual case study. The aim is to communicate without losing the perspective either of the approaches advances. Thus, a step forward towards an integrative vulnerability assessment might be to strengthen the dialogue between different groups of experts in natural hazard vulnerability/risk assessment through exchange of views about definitions, concept and underlying worldviews and values [60].

In terms of vulnerability/risk assessment outcomes, there are three main types of methods (results) - quantitative, semi-quantitative and qualitative, all with benefits and drawbacks. The main difference between quantitative and qualitative methods lies in the fact that quantitative assessments provide a more explicit objective framework which may be conducive to improving decision making process. However, the most appropriate tool depends on the decision problem at hand (for example, qualitative vulnerability assessment can be more cost effective, less time consuming and easier to understand for non-technical stakeholders), the objective (including scale) of the analysis and the quality/quantity of available data. Hence there is no general preference for qualitative, semi-quantitative or quantitative approaches [61]. One must also acknowledge that there is no quantitative vulnerability/risk assessment totally devoid of expert judgment; quantitative vulnerability/risk analysis rather provides a framework for making systematic judgment [62]. It is the quality and quantity of subjectivity that affects the overall outcome of the analysis.

With regards to uncertainty in vulnerability analysis, Gall [63] emphasizes the importance of knowledge quality assessment - 'uncertainty and sensitivity analysis are mandatory for maximizing methodological transparency and soundness, and hence the acceptance of research findings; despite this demand, both analyses are often missing in vulnerability assessment'. However, progress has been done, for example, in the field of technical (structural) vulnerability (mostly, for hazards like floods and earthquakes), where empirical as well as statistical (probabilistic) methods aided by GIS and advanced computational models are used to estimate uncertainty in vulnerability and its components.

To allow for an improved decision making process through the treatment of uncertainty, first the joint effort between end-users and experts must shift towards a more transparent, participative and open process. The role of the scientist seen as 'speaking truth to power' is defective as it implies that all uncertainties can be treated. Conversely, experts should clearly communicate the limitations of their findings as well as continue to investigate the effects of uncertainty on risk and its determinants in order support the community to face future challenges in dealing with natural hazards and risk and global change.

Notes

- Coping capacity is the ability of people, organizations and systems, using available skills and resources, to face and manage adverse conditions, emergencies or disasters [8]

- Capacity is the combination of all the strengths attributes and resources available within a community, society or organization that can be used to achieve agreed goals [8]

- Exposure is defined as the totality of people, property, systems or other elements present in hazard zones that are thereby subject to potential losses [8]

- Resilience is the ability of a system, community or society exposed to hazards to resist, absorb, accommodate to and recover from the effects of a hazard in timely and efficient manner, including through the preservation and restoration of its essential basic structures and functions [8]

- Stress is a continuous or slowly increasing pressure, commonly within the range of normal variability. Stress often originates and stressors (the sources of stress) often reside within the system [21]

ACKNOWLEDGEMENTS

This study was prepared in the frame of the research project Changing Hydro-meteorological Risks as Analyzed by a New Generation of European Scientists (CHANGES), a Marie Curie Initial Training Network, funded by the European Community's 7th Framework Programme FP7/2007-2013 under Grant Agreement No. 263953.

REFERENCES

1. Facebook Twitter Bibsonomy   CiteULike   Reddit   LinkedIn StumbleUpon Mail to a Friend

2. CRED EM-DAT. The International Disaster Database. http://www.emdat.be/accessed 20 August 2012

3. UNISDR. Global Assessment Report on Disaster Risk Reduction 2011http://www.preventionweb.net/english/hyogo/gar/2011/en/home/index.htmlaccessed 20 August 2012).

4. D Alexander, Vulnerability to landslides. In: Glade T, Anderson MG, Crozier MJ, (eds.). Landslide Hazard and Risk. Chichester: John Wiley & Sons; 2005175198

5. World Bank. Natural Disaster Risk Management in the Philippines: Enhancing Poverty Alleviation through Disaster Reduction   2005http://openknowledge.worldbank.org/handle/10986/8748accessed 20 August 2012).

6. A Farrell, S. D Vandeveer, J Jager, Environmental assessments: four under-appreciated elements of design. Global Environmental Change 2001114311333

7. D Schröter, W Cramer, R Leemans, I. C Prentice, M. B Araújo, N. W Arnell, et alEcosystem Service Supply and Vulnerability to Global Change in Europe. Science 2005310575213331337

8. D Schröter, Vulnerability to changes in ecosystem services. CID Graduate Student and Postdoctoral Fellow Working Paper 102005accessed.

9.   UNISDR. Living with Risk. A global review of disaster reduction initiatives: United Nations; 2004http://www.unisdr.org/files/657_lwr21.pdfaccessed 20 August 2012).

10.   UNISDR. Terminology on Disaster Risk Reduction. Geneva, Switzerland   2009http://www.unisdr.org/files/7817_UNISDRTerminologyEnglish.pdfaccessed 21 August 2012).

11.   F Nadim, H Einstein, W Roberds, Probabilistic stability analysis for individual slopes in soil and rock State of the art Paper 3. Proceeding of the Int Conf on Landslide Risk Management, 31 May- 2 June, 2005Vancouver, Canada 2005.

12.   J Birkmann, Indicators and criteria for measuring vulnerability: theoretical basis and requirements. In: Birkmann J, (ed.). Measuring vulnerability to natural hazards Towards disaster resilient societies. Tokyo: United Nations University; 20065577

13.   N Brooks, Vulnerability, Risk and Adaptation: A Conceptual Framework 20032003120http://www.gsdrc.org/go/display&type=Document&id=3979accessed 22 August 2012).

14.   P Blaikie, at risk: natural hazards, people's vulnerability, and disasters. London; New York: Routledge; 1994

15.   UNDRO. Natural disasters and vulnerability analysis: report of Expert Group Meeting. Geneva: Office of the United Nations Disaster Relief Co-ordinator; 1979http://www23.us.archive.org/details/naturaldisasters00offiaccessed 25 August 2012).

16.   S. H Schneider, S Semenov, A Patwardhan, I Burton, Magadza CHD, Oppenheimer M, et al. Assessing key vulnerabilities and the risk from climate change. Climate Change 2007Impacts, Adaptation and Vulnerability. Contribution of Working Group II to the Fourth Assessment Report of the Intergovernmental Panel on Climate Change. In: Parry ML, Canziani OF, Palutikof JP, van der Linden PJ, Hanson CE, (eds.). Cambridge, UK; 2007. 779810

17.   H-M Füssel, R. T Klein, Climate Change Vulnerability Assessments: An Evolution of Conceptual Thinking. Climatic Change 2006753301329

18.   C Vogel, O Brien, K Vulnerability, and global environmental change: rhetoric and reality. Aviso: An Information Bulletin on Global Environmental Change and Human Security. 2004200418

19. H-G Bohle, Vulnerability and Criticality: Perspectives from Social Geography. IHDP Update 2/2001, Newsletter of the International Human Dimensions Programme on Global Environmental Change. 2001200117

20. R Davidson, An Urban Earthquake Disaster Risk Index. The John A. Blume Earthquake Engineering Center, Department of Civil Engineering, Stanford University, 1997Report 121

21. C Bollin, C Cardenas, H. H Vatsa, KS. Natural Disaster Network; Disaster Risk Management by Communities and Local Governments. Washington, D.C.: Inter-American Development Bank, 2003

22. B. L Turner, R. E Kasperson, P. A Matson, J. J Mccarthy, L Corell, R. W Christensen, L, et al. A framework for vulnerability analysis in sustainability science. Proceedings of the National Academy of Sciences of the United States of America, 2003

23. B Wisner, P Blaikie, T Cannon, I Davis, Natural hazards, people's vulnerability and disasters, (2nd ed.). London- New York: Routledge; 2004

24. M-L Carreño, O Cardona, A Barbat, Urban Seismic Risk Evaluation: A Holistic Approach. Natural Hazards 2007401137172

25. T Glade, Vulnerability assessment in landslide risk analysis. Die Erde 20031342121138

26. S. L Cutter, Vulnerability to environmental hazards. Progress in Human Geography 1996204529539

27. M Papathoma-köhle, M Kappes, M Keiler, T Glade, Physical vulnerability assessment for alpine hazards: state of the art and future needs. Natural Hazards 2011582645680

28. G Hufschmidt, M Crozier, T Glade, Evolution of natural risk: research framework and perspectives. Natural Hazards and Earth System Sciences 200553375387

29. Van Westen, CJ, Van Asch TWJ. Landslide hazard and risk zonation-why is it still so difficult? Bulletin of Engineering Geology and the Environment 2006652167184

30. Consortium CapHaz-Net. Social Capacity Building for Natural Hazards Toward More Resilient Societies. http://caphaz-net.org/project-overviewaccessed 25 August 2012

31. UNDP. United Nations Development Program's Human Development Index (HDI). http://hdr.undp.org/en/statistics/hdi/ accessed 26 August 2012

32. S. L Cutter, B. J Boruff, W. L Shirley, Social Vulnerability to Environmental Hazards. Social Science Quarterly 2003842242261

33. Consortium CapHaz-Net. Social Capacity Building for Natural Hazards Toward More Resilient Societies, Social vulnerability to Natural Hazards, Deliverable D4.1, CapHaz-Net Project. 2010

34. V. W Maclaren, Urban Sustainability Reporting. Journal of the American Planning Association 1996622184202

35. Consortium MOVE. Methods for the Improvement of Vulnerability Assessment in Europe, Guidelines for development of different methods, Deliverable D6, MOVE Project. 2010

36. M. S Kappes, M Papathoma-köhle, M Keiler, Assessing physical vulnerability for multi-hazards using an indicator-based methodology. Applied Geography 2012322577590

37. H Apel, A. H Thieken, B Merz, G Blöschl, Flood risk assessment and associated uncertainty. Nat Hazards Earth Syst Sci 200442295308

38. I Kelman, R Spence, Flood Failure Flowchart for Buildings. Proceedings of the Institution of Civil Engineers-Municipal Engineer, 2003

39. Consortium ENSURE. Enhancing resilience of communities and territories facing natural and na-tech hazards, Methodologies to assess vulnerability of structural systems Del1.1.1, ENSURE Project. 2009

40. B Merz, H Kreibich, U Lall, What are the important flood damage-influencing parameters?A data mining approach. EGU General Assembly 2012

41. B Büchele, H Kreibich, A Kron, A Thieken, J Ihringer, P Oberle, et alFlood-risk mapping: contributions towards an enhanced assessment of extreme events and associated risks. Nat Hazards Earth Syst Sci 200664485503

42. University of Technology Hamburg TFlood Manager E-learning. http://daad.wb.tu-harburg.de/?id=1399accessed 28 August 2012

43. FEMA Hazus, U.S. Multi-Hazards Flood Model. http://www.fema.gov/hazusaccessed 28 August 2012

44. P. J Finlay, R Fell, Landslides: risk perception and acceptance. Canadian Geotechnical Journal 1997342169188

45. T Glade, M. G Anderson, M. J Crozier, Landslide Hazard and Risk: John Wiley & Sons; 2005

46. M Uzielli, Vulnerability and risk assessment for geohazards. Probabilistic estimation of regional vulnerability to landslides. International Centre for Geohazards (ICG), Norwegian Geotechnical Institute (NGI), 2007

47. R Fell, Landslide risk assessment and acceptable risk. Canadian Geotechnical Journal 1994312261272

48. M Cardinali, P Reichenbach, F Guzzetti, F Ardizzone, G Antonini, M Galli, et alA geomorphological approach to the estimation of landslide hazards and risks in Umbria, Central Italy. Nat Hazards Earth Syst Sci 2002

49. Maquaire, C Weber, Y Thiery, A Puissant, J-P Malet, A Wania, Current practices and assessment tools of landslide vulnerability in mountainous basins. Identification of exposed elements with a semi-automatic procedure. Landslides Evaluation and Stabilization, Proceedings of the 9th International Symposium on Landslides, 2004Rio de Janeiro, Brazil: Balkema, Rotterdam.

50. F Leone, J. P Asté, E Leroi, Vulnerability assessment of elements exposed to mass movement: Working toward a better risk perception. In: Senneset K, (ed.). Landslides- Glissements de Terrain, Rotterdam: Balkema; 1996

51. F. C Dai, C. F Lee, Y. Y Ngai, Landslide risk assessment and management: an overview. Engineering Geology 20026416587

52. M Uzielli, F Nadim, S Lacasse, A. M Kaynia, A conceptual framework for quantitative estimation of physical vulnerability to landslides. Engineering Geology 2008

53. Z Li, F Nadim, H Huang, M Uzielli, S Lacasse, Quantitative vulnerability estimation for scenario-based landslide hazards. Landslides 201072125134

54. S Fuchs, K Heiss, J Hübl, towards an empirical vulnerability function for use in debris flow risk assessment. Nat Hazards Earth Syst Sci 200775495506

55. M Kaynia, M Papathoma-köhle, B Neuhäuser, K Ratzinger, H Wenzel, Z Medina-cetina, Probabilistic assessment of vulnerability to landslide: Application to the village of Lichtenstein, Baden-Württemberg, Germany. Engineering Geology 2008

56. M. G Winter, J. T Smith, S Fotopoulou, K Pitlakis, O-C Mavrouli, J Corominas, et alDetermining the physical vulnerability of roads to debris flow by means of an expert judgement approach. EGU General Assembly 2227April, 2012Vienna, Austria; 2012.

57. M Uzielli, S Duzgun, B. V Vangelsten, A First-Order Second Moment framework for probabilistic estimation of vulnerability to landslides. Proceedings ECI Conference on Geohazards, Lillehammer, Norway, 1821June 2006

58. M. H Jahanpeyma, Parvinnezhad Hokmabadi D, Rahmanizadeh A. Analytical Evaluation of Uncertainty Propagation in Seismic Vulnerability Assessment of Tehran Using GIS. Journal Civil Eng Urban 2011110509http://www.ojceu.ir/mainaccessed 28 August 2012).

59. F Berkes, Understanding uncertainty and reducing vulnerability: lessons from resilience thinking. Natural Hazards 2007412283295

60. S Dessai, M Hulme, Does climate adaptation policy need probabilities? Climate Policy 200442107128

61. S Fuchs, C Kuhlicke, V Meyer, Editorial for the special issue: vulnerability to natural hazards-the challenge of integration. Natural Hazards 2011582609619

62. Hufschmidt, G, Glade, T, Vulnerability analysis in geomorphic risk assessment. In: Alcántara-Ayala, I, & Goudie, A. S. (eds.). Geomorphological Hazards and Disaster Prevention: Cambridge University Press; (2010)., 233-243.

63. K Ho, E Leroi, W. R. Quantitative risk assessment- application, myths and future direction, Keynote Paper. Proceedings of the International Conference on Geotechnical Engineering (GeoEng 2000), November 2000Melbourne, Australia.

64. M Gall, Indices of social vulnerability to natural hazards: a comparative evaluation, PhD Thesis. Columbia City: University of South Carolina; 2007

The Role of Earthquake Information Management System to Reduce Destruction in Disasters with Earthquake Approach

Sima Ajami[1]

[1]Department of Health Information Technology, Health Management and Economics Research Centre, School of Medical Management and Information Sciences, Isfahan University of Medical Sciences, Isfahan, Iran

INTRODUCTION

Iran, because of extent, geographical situation and climatic variety, is one of the disaster-prone countries of the world. [1] Natural disasters, for example earthquake, are an unexpected event that cause damage and destruction to human life and health, and the injured persons without others assistance are not able to meet their need. Earthquakes in Iran and neighboring regions (e.g., India, Turkey and Afghanistan) are closely connected to their position within the geologically active Alpine-Himalayan belt (Table 1). [2-5] Earthquake crises disrupt all daily affairs of society, such as economic activities, city services, communication systems and community services and public health. [6] An Earthquake Information Management System (EIMS) is a system that records, collects, keeps, retrieves and analyzes inputs, produces reports and required earthquake information (EI) and renders them to the right people and organizations to manage earthquake response activities. [7] EI is not an end in itself, but it helps to make better decisions in designing policies, response planning, management of disasters, monitoring and evaluating disaster programs and services, and reducing damages. [8]

Table 1: Deadliest earthquakes by year, 1995–2005

Deadliest earthquake				
Year	**Date**	**Magnitude**	**Fatalities**	**Region**
2005	03/28	8.7	1313	Northern Sumatra, Indonesia
2004	12/26	9.0	283,106	Off the west coast of Northern Sumatra
2003	12/26	6.6	31,000	Southeastern Iran
2002	03/25	6.1	1000	Hindu Kush region, Afghanistan
2001	01/26	7.7	20,023	India
2000	06/04	7.9	103	Southern Sumatera, Indonesia
1999	08/17	7.6	17,118	Turkey
1998	05/30	6.6	4000	Afghanistan–Tajikistan border region
1997	05/10	7.3	1572	Northern Iran
1996	02/03	6.6	322	Yunnan, China
1995	01/16	6.9	5530	Kobe, Japan

PROBLEM STATEMENT

Unfortunately, information systems in most countries are inadequate to provide the needed management support. Earthquake loss estimates are forecasts of damage as well as human and economics impacts that may result from future earthquakes. These estimates are based on current scientific and engineering knowledge. [9] The "earthquake loss estimation methodology" is a system that uses mathematical formulae and information about building stock, local geology and the location and size of earthquake risk, economic data and other information to estimate losses from a potential earthquake. The EIMS uses ArcGIS (geographical information system) to map and display ground shaking, the pattern of building damage and demographic information about a community. Once the location and size of a hypothetical earthquake are established, EIMS will estimate the distribution of the amounts of the following: ground shaking, buildings damaged, injured persons, damage to transportation systems, disruption of electrical and water utilities, displacement of populations and cost of repair of likely damage. [10-13]

Estimation of losses from future earthquakes is essential for preparation for disasters and it facilitates better decision making at local, regional, provincial and national levels of government. An EIMS can estimate earthquake losses to support land-use planning and facility site decisions (e.g., a map-based analysis of the potential intensity of ground shaking from a postulated earthquake that identifies those parts of the community that will experience the most violent shaking and the buildings at greatest risk of damage); prioritization of retrofit or abatement programs (e.g., an estimate of building damage that provides the basis for establishing programs to mitigate or strengthen buildings that may collapse in earthquakes by providing estimates of damages and casualties); regional, provincial and local emergency response and contingency planning (e.g., estimates of casualties and of damage to buildings and utilities); medical and relief agency preparedness and response (e.g., estimates of casualties and homelessness); and assistance planning. [11]

In this research, I seek to answer five important questions: When are EIMS useful? What are the essential substructures in an EIMS? What are the important functions of an EIMS? What are the steps taken to create an EIMS? and How is an EIMS used to prepare for earthquakes?

AIM

In this study, the information management networks related to earthquakes in India, Afghanistan, Japan, and Turkey were compared to Iran to rationally determine the relative strength of the EIMS in each country. Their weaknesses and strengths were determined. And several recommendations and a model were developed to eliminate weaknesses, improve the efficiency of EIMS, reduce damages and losses and expedite relief to victims after earthquakes.

METHODOLOGY

This research is empirical and the study was an analytical comparison. The data consisted of the population of EIMS employed in India, Afghanistan, Japan, and Turkey and Iran. These countries were chosen because they are among the countries with the most experience with extreme events in Asia, and they all experience devastating earthquakes (Table 1).

To perform this study, I developed forms and questionnaires for data collection through interviews and observations. The forms were developed to define standard characteristics of an information management system by extracting guidelines from the Joint Commission on Accreditation of Healthcare Organization (JCAHO), the American Health Information Management Association (AHIMA) and Canadian Council on Health Services Accreditation (CCHSA) [14] and synthesizing the information. The questionnaire was designed to acquire the opinions of experts to enable the weighting of each characteristic of an EIMS. In the first phase of data collection, the forms were validated and the questionnaire was approved. Internet sources, professional personnel, documents, journals and books were consulted to develop the data which included EI sources, methods of recording, storing, retrieving, analyzing, interpreting, and distributing EI, national and international usage of EI, and so on. The Criteria Rating Technique [15] and the descriptive method were used to analyze findings. Standard characteristics of information management systems were selected as criteria.

To compare the importance of the characteristics of EIMS, experts were asked to set weights (from 1, of low importance, to 10, of high importance) by brainstorm. The means of the experts' opinions of the weight of each criterion were calculated (Table 2 and 3). Ratings were established (ratio = weight of each criteria divided by sum) and scales (positive = 4, moderate = 3, not access = 2, negative = 1) and scores (score = ratio*scale) for selected countries were calculated.

Table 2: The EIMS characteristics evaluating in selected countries

Criteria/country	Weight	Ratio	India		Afghanistan		Iran	
			Scale	Score	Scale	Score	Scale	Score
1) Information sources are existed	8	0.11	4	0.44	4	0.44	3	0.33
2) Users of EI are specified	6	0.08	4	0.32	3	0.24	3	0.24
3) System has security process	6	0.08	4	0.32	2	0.16	2	0.16
4) EI is recorded and stored systematically	7	0.09	4	0.36	4	0.36	1	0.09
5) EI is retrieved, analyzed and interpreted systematically	9	0.12	4	0.48	2	0.24	3	0.36
6) Administrators are specified in various functions	9	0.12	4	0.48	4	0.48	3	0.36
7) No parallel and repeated activities by various organizations	5	0.07	4	0.28	4	0.28	1	0.07
8) EI is distributed and used in national and international levels	6	0.08	4	0.32	4	0.32	3	0.24
9) EIMS has feedback	9	0.12	4	0.48	4	0.48	3	0.36
10) Accessibility of EI is easy and fast	8	0.11	4	0.44	3	0.33	1	0.11
Sum	73	1		4		3.3		2.32

Table 3: The EIMS characteristics evaluating in selected countries (continued)

Criteria/country	Weight	Ratio	Japan		Turkey		Iran	
			Scale	Score	Scale	Score	Scale	Score
1) Information sources are exist ed.	8	0.11	4	0.44	4	0.44	3	0.33
2) Users of E.I. are specified.	6	0.08	4	0.33	4	0.33	3	0.25
3) System has security process.	6	0.08	2	0.16	4	0.33	2	0.16
4) E.I. is recorded and stored systematically	7	0.09	4	0.38	4	0.38	1	0.09
5) E.I. is retrieved, analyzed and interpreted systematically.	9	0.12	4	0.49	4	0.49	3	0.37
6) Administrators are specified in various functions.	9	0.12	4	0.49	4	0.49	3	0.37
7) No parallel and repeated activities by various organizations.	5	0.07	4	0.27	4	0.27	1	0.07

		Ratio	Scale	Score	Scale	Score	Scale	Score	Scale	Score
8) E.I. is distributed and used in national & international levels.	6	0.08	4	0.33	4	0.33	3	0.25		
9) EIMS has feedback.	9	0.12	4	0.44	4	0.44	3	0.49	3	0.37
10) Accessibility of E.I. is easy and fast.	8	0.11	4	0.44	4	0.44	1	0.11		
Sum	73	1		3.77		4		2.37		

[i].　Earthquake Information= EI; Ratio = weight of each criteria/sum; Score = ratio*scale

[ii].　Scales: Positive = 4, moderate = 3, negative = 2, not access = 1

[iii].　Range of ranks: 1–1.6: Very weak; 1.7–2.2: Weak; 2.3–2.8: Moderate; 2.9–3.4: Good; 3.5–4: Very good

RESULTS

The results of the data collection provide answers to the five research questions. Each answer is described briefly below.

When are EIMS useful? There are four conditions that make EIMS more useful: when information users and addressers are specified; when time, form and the mechanism of information distribution are specified; when the EI is valid and reliable; and when there is fast access to EI. Furthermore, the data received are often not helpful for management decision making because they are incomplete, inaccurate, untimely and unrelated to the priority tasks and functions of crisis management.

What are the essential sub-structures of an EIMS? The information compiled and surveys indicate that the essential sub-structures that must be addressed or included in an EIMS are the nature of: crisis management, information technology, the geographic information system, the earthquake information system, mass media communication, cell phone communication, capital and human resources.

What are the important functions of an EIMS? This study finds that EIMS are needed: for fast and easy retrieval of information, which is very difficult; for extraction and access of information for managers and related users; for integrating data from different sources; for reduction of parallel and redundant activities by responding organizations; to decrease cost and time; for assessment and monitoring of plans before and after earthquakes; to identify training and function needs; and to formulate prevention, action and rehabilitation actions.

What are the steps taken to create an EIMS? The following steps are usually taken: establish a joint commission of governmental and non-governmental sectors and organizations; determine the primary participants in earthquake management; determine and formulate a plan for a system based on general principles and goals; identify the data needed for the system; identify the sources for the system's informatics; identify registration, collection and storage methods of and administrators for the system; determine the retrieval and analytical methods of and administrators for the system; establish the information methods and distribute them to the administrators of the system; establish methods of systematic communication among the administrators; create a mechanism to render feedback for system improvement; and ensure that the system and its plans and functions are dynamic and flexible.

How is an EIMS used to prepare for earthquakes? The first step in preparing for a disaster is estimate and assess its potential impact. These estimations and assessment can provide the basis for developing mitigation policy, developing and testing emergency preparedness and responsing for post-disaster, and reliefing negative outcomes. [16] Reducing earthquake loss begins before the earthquake. Loss estimates provide public and private sector agencies with a basis for planning, zoning, building codes and development regulations, and policy that would reduce the risk posed by violent ground shaking and ground failure. Loss estimates can also be used to evaluate the cost-effectiveness of alternative approaches to strengthening potentially hazardous structures. Preparing to respond, understanding the scope and complexity of earthquake damage is essential to effective preparedness. The EIMS can forecast damage to buildings, casualties and disruption of utilities. These estimates can serve as the basis for developing emergency response plans and for organizing tests and exercises of response capability.

ABOUT EIMS

Beginning in the late 1950s, planners began development and use of computerized models, planning information systems and decision-support systems to improve EI management performance. They have found tools to enhance their analytical and geospatial technologies which may be different from one country to another. The industrialized nations are well adapted to this information technology. They use it in many fields. Governments apply urban information systems in all aspects of the planning process, including data collection, storage, data analysis and presentation, planning and policymaking, communication with the public, policy implementation and administration. The United States is the pioneer in this field. They began working with urban information systems in the 1970s. Canada and Australia have developed systems. And European countries like France, Germany and the Netherlands have been successful in applying these technologies. Turkey is a latecomer in this field because of financial constraints, their other priorities, a lack of technical expertise and different administrative mentalities. But today, the urban information system is a popular notion among local governments in Turkey. The first initiative of local governments to use urban information systems was in the cities Bursa and Ayden beginning in the mid-1990s. Since then three other metropolitan municipalities, Istanbul, Ankara and Izmir, studied the digitization of maps and plans and began to create inventories of their cities.

In India, findings showed that the Disaster Information Management System (DIMS) was launched by SRISTI. The SRISTI participated in relief and rehabilitation work in Kutch. However, their relief work suffered immensely due to lack of information and proper planning. When answers to important questions that were cropping up were needed – for instance, whether there is a database on the distribution of available resources and expertise with individuals, institutions and corporations – SRISTI discovered the information wasn't available. This revealed the need for a system for disaster mitigation and for documenting the experiences of individuals and organizations, which might provide a knowledge database that can assist coordination in future disasters. Thus, SRISTI initiated an effort to build a "Disaster Management Information System." Through this initiative, the development of a database-driven information system for Disaster Management Authorities (DMA) in various states, NGOs and other organizations is underway. SRISTI appealed to NGOs, relief workers, DMAs and individuals to share their experiences and volunteer their services and resources to the online database maintained on the SRISTI website. The database currently contains information from more than a thousand volunteers who have offered their services and resources in times of emergency. About 700 organizations and institutions are indexed on the site, as are other resources and web links. The DMIS is a voluntary activity run with in kind contributions of time and services by SRISTI volunteers, NGOs and, above all, civil institutions across the world. All the information shared with us is accessible to all, except in cases in which the volunteer has chosen to limit access to the relevant authorities. [17-18]

EIMS IN AFGHANISTAN

Our research reinforces the belief that Afghanistan possesses the potential for many natural disasters. Therefore, a DIMS, especially an EIMS, would be particularly helpful. Disaster management is a needs a wide variety of information, needs to track different locations and during different periods, and this information must have a set format in order that key staff can employ the information in decisions. A new project named "Management Information for Natural Disasters" (MIND) in Kabul and Kunduz provinces in Afghanistan has been operating since the beginning of 2012 and has been doing well for the last 8 months. Its goal was to establish and expand a crisis information management system. It is frequently updated and provides information to governments. MIND has increased Afghanistan's crisis management capacity nationally, supporting the training of disaster managers and improving city services by building governmental organizations in management information for natural disasters. Natural disaster management in these two provinces is primarily guided by estimates of damages which are used to direct rescue operations. There is currently no system to avoid or decrease disaster damages. Before and after a disaster, management is very weak in these countries and they are therefore dependent on the UN or NGOs for response. Information needed for the EIMS includes the availability and distribution of response personnel and spatial information stored in GIS to determine the areas of greatest destruction and locations of great danger. In this system, satellite data are crucial for identification of crisis locations and understanding their distribution. [17,19] To record the information that is acquired, an Information System Unit is trained to input data and information and to manage stations of the informative system. The DIMS for earthquakes has tended to not be useful in many parts of the country primarily because of the lack of timely information from disaster areas.

EIMS IN JAPAN

The Japanese DIMS is called "PHOENIX" (Preement Hyogos Emergency Management Network for Disaster Information Exchange). [20] In this system, information about the amount and degree of earthquake damages and on-going developments of conditions in a disaster area are collected and processed. Information is provided by agencies, involved organizations, individuals (including those from governments, experts, volunteers and others), monitoring stations and other data sources employing several communication methods (including the

Internet, radio, print, televison and satellite). Recording and collecting earthquake information is centralize by informatics centers of local society of province. Japan's Red Cross, Rescue Team, NGOs, health organizations and others [21] do the estimating, calculating, and publishing of the information that is needed, and they send them to the relevant agencies. PHOENIX crisis management is designed for access and use by individuals, local communities and the national government. All satellite, land-based and atmospheric data will be transmitted over the Internet and displayed on local pages for Hyogo people. This system has been established for the entirety of Japan. [20] The Hygo Province crisis management network is directly supervised by the prime minister and his/her ministry. According to their documents, this system has been very successful. So it is expected that after they have completed their analysis, the system will be on-line for local, national and international users. [20]

EIMS IN TURKEY

Our findings reveals that every governmental office in Turkey has begun development of electronic systems to meet their needs. Hence, e-government has become an important tool. As a result, the Turkish government has resolved to provide public services online in accordance with EU targets. As a part of this process, the prime ministry of Turkey has chosen Istanbul as a pilot-project area where many complex governmental tasks are carried out. Turkey is situated in an extremely active seismic zone. Since Istanbul has been growing rapidly without proper planning, great precautions should be taken to mitigate and prepare for future disasters. Therefore, Turkey has committed to build a natural disaster management information system immediately and it is called AFAYBIS. AFAYBIS is designed as a minor part of their e-Government system. It was based upon the information acquired in surveyes of government and private-sector data providers. The system will use geographical analysis to identify the regions with the greatest disaster potential. The project is also intended to quickly and effectively create a tool for management of response and relief during and after disasters. After an analysis of the current state of affairs was completed, the data and the data sources were identified. The system is designed as two parts: a database and a communication system.

The communications component is to constantly update data before disasters and to provide continuous supply of data. Consequently this disaster information system is designed according to the standards of the Turkish e-government, which is always intended to be up-to-date. The disaster management information system that has been developed to solve this problem will result in optimum efficiency during disaster. This system contains the structure that determines its relationships to data and access to its information, disaster management communication, risk mitigation and disaster preparation, and post-disaster coordination of the prime minister, governors and other institutions. The services and duties of the institutions are also developed into the system so as to avoid modification of their existent organizational structures. [22] Earthquake management in Turkey, without exception, is a problem. It is unsystematic, unplanned, static and awareness of it is low.

EIMS IN IRAN

Disaster management systems are often designed because of the lack local management. Disasters are managed nationally in Iran and such management tends to cause disasters to spread, impacting many more people. [1] Despite the national approach, there is no official department called the "Department of Earthquake Information Management System." There is an EIMS. Information has been recorded by hand or using computers. Management of disasters has been done through the crisis office of the health ministry. All of the universities of Iran and the health minister are responsible for information management. This system is equipped to communicate information from crisis offices in the country's universities. After an earthquake, these units are prepared to produce up-to-date information (about hospitals, the Red Crescent Society, ambulances, facilities and other important resources) and report it to the crisis-control office. In universities, the crisis rooms are expected to report information as rapidly as possible. The EIMS in Iran tended to be incapable of reporting important information that was needed in advance of, during and after earthquakes. Defective, insufficient and inaccurate registration of data, declaration and publication of different and contradictory population data and a lack of reliable information disabled the development of preventative systemic planning. To make an EIMS work in Iran, we need to provide support for managers, and to do this, modified model of the EIMS should be designed. [1] The modified model includes information about: the responsible organizations, their functions, and the work-flow that were reflected in the Delphi Technique.

We offer a model that shows the relationships between organizations related to the EIMS in Iran (Figure 1). In this model, the responsibility and function of every organization is determined. These duties are classified according to registration and collection of earthquake data, the storage and processing of these data, and the analysis and distribution of information and recommendations produced by the EIMS.

EVALUATING THE EFFECTIVENESS OF EIMS DESIGN IN EACH COUNTRY

The highest sum of scores of the effectiveness of EIMS are found in India and the lowest in Iran (Table 2 and 3). The weaknesses of EIMS are found in that: the EI stores systematically, there are parallel and repeated activities by various organizations, and the access to the EI is not easy and nor fast, particularly in Iran's EIMS. In the range of ranks, Afghanistan's and India's systems were classified in the very good range and Iran's was in the moderate range. [17]

The use of mass media is imperative for communicating news and information to the public. Responsible journalism can also help to clear up inaccurate rumors and to influence the public's attitude toward preparing for disasters. Moreover, press coverage of old disasters may provide good data to fill the gaps in circumstances where official records do not exist. However, this study has revealed that the press has largely failed in terms of guidance toward disaster mitigation and preparedness. It seems that media are more interested in reporting disastrous news than informing the public of ways to avoid disasters. The Turkish media has been more influential in urging both the public and government officials to prepare for the earthquakes. Television appears to be a more effective tool to achieve this. [23]

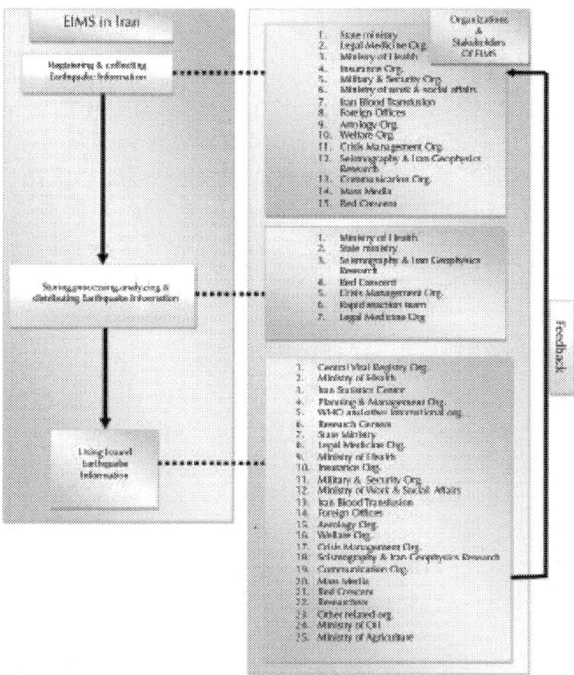

Figure 1: The proposed model of EIMS that shows process of relationships between organizations related to EIMS in Iran.

The Japan Meteorological Agency (JMA) is responsible for producing EI and tsunami forecasts and they have developed an earthquake notification system in Japan. At present, JMA issues the following kinds of information successively when a large earthquake occurs: prompt reports of the occurrence of large earthquakes and major seismic intensities caused by earthquakes within about 2 minutes after the earthquake; tsunami forecasts in around 3 minutes; expected arrival times and maximum heights of tsunami waves in around 5 minutes; and location of the hypocenter and the magnitude of the earthquake, intensity at each observation station, times of high tides and the expected tsunami arrival times within 5–7 minutes. To issue the above information, JMA has established an advanced nationwide seismic network with about 180 stations for seismic-wave observation and about 3400 stations for instrumental seismic intensity observation, in addition to the approximately 2800 seismic intensity stations maintained by local governments. [24]

CONCLUSIONS

The initial effort to systematically collect and analyze data in developing countries should be undertaken by national program managers. Based on our investigation of current earthquake information management in India, Afghanistan and Iran, we can stress the need for further development of EIMS because of: the critical need for the information that must be gathered; a concern about continuous improvement of the data made available; its ability to help manage emergency response and relief in natural disasters (through more rapid availability and retrieval in an EI); a need for timely reporting and feedback to managers; the need to analyze information and render reports that define strengths, weaknesses, threats and opportunities; the need for monitoring the status of healthcare services and the needs of such services; the crucial need to coordinate activities between government and non-governmental sectors through the EIMS; the need to reduce deaths and establish health priorities and planning to decrease mortality after future earthquakes; the ability to use the outcomes of crisis to determine the causes of earthquake mortalities and other problems in order to prevent future impacts; and the need to formulate strategies for disease prevention and to reduce preventable deaths in earthquake zones.

A rapid response to a damaging earthquake will reduce the loss of life, lessen the complications that stem from injuries and secondary damage, and expedite relief to victims. Reliable and up-to-date information can have an impact on the mitigation of risk and prevention of disaster. Because of the financial and human costs of disasters, the establishment of a general, scientific and practical earthquake information management network is desperately needed.

ACKNOWLEDGMENTS

The author would like to thank Misses Z. Moradi, Mahshid Fattahi and N. Nematolahi for helping to fulfill this research.

REFERENCES

1. Bamdad N. The Role of Community Knowledge in Disaster Management: The Bam Earthquake Lesson in Iran. Institute of Management and Planning Studies, Tehran, Iran. Available from: http://www.engagingcommunities2005.org/abstracts/S93-bamdad-n.html/(accessed 11 September 2012).

2. Tavakoli N. Health information management in disaster' proceeding of second national health and management disaster meeting, 2004 Nov 24–26, Tehran, Iran.

3. Rastegari H, Ajami S. An overview of the management of a crisis. Health Information Management 2005;2(3):73-81. http://him.mui.ac.ir/index.php/him/article/view/28/26.

4. Ajami S. The role of information management in rendering healthcare in disasters. In Proceeding of Second National Health and Management Disaster meeting, 2004 Nov 24–26, Tehran, Iran.

5. The U.S. Geological Survey, Earthquake Hazards Program. Disaster Management Information System, fact sheet 04: Largest and Deadliest Earthquakes by Year 1990 – 2005'. Available from: http://www.sristi.org/dmis/facts.(accessed in 2009).

6. Division of International Health, Department of Health and Human Services, Centers for Disease Control and Prevention, Epidemiology Program Office. Public Health Systems Development; Health Information Systems. [online]. United States. Available from: http:// www.cdc.gov/epo/dih/systems.html.(accessed in 2000).

7. Seismological Bureau of Yunnan Province. Improvement of Earthquake disaster reduction and early warning systems. [online]. PR, China. Available from: http://www.chinaproject.network. (accessed in 2001).

8. Lippeveld T, Sauerborn R, Bodart C. Design and implementation of health information systems. Geneva, Switzerland: World Health Organization; 2000.

9. Rock ML. Effective crisis management planning: Creating a collaborative framework [online]. Educ Treat Child J 2000;23(3):248-64.

10. Seeger MW, Sellnow TL, Ulmer RR. Communication and organizational crisis. USA:Praeger; (accessed 11 April 2006).

11. Kabirzadeh A. Disease control and prevention after natural disaster. In Proceeding of Natural Health and Management of Disaster Meeting, 2003 May 27-29, Tehran, Iran. In Persian

12. Ajami S, Fatahi M, Moradi Z, Nematolahi N. An analysis studies on Earthquake Information Management Systems (EIMS) in Japan, Afghanistan and Iran and proposing a suitable model for Iran. In Proceeding of International Disaster Reduction Conference (IDRC), From 27 August to first of September 2006, Switzerland, Davos.

13. Ajami S, Fatahi M, Moradi Z. Reduce destroys and rule of Earthquake Information Systems the Comparative study in Turkey, Afghanistan and Iran. In Proceeding of International Disaster and Risk Conference IDRC, from August 25 to August 29, 2008, Switzerland, Davos.

14. Ajami S, Tavakoli-Moghadam O. A Comparative Study on Health Information Management System with Standards in Ayatolah-Kashani Hospital in Isfahan, Iran. Health Information Management; 2006;1(3): 63-7. http://him.mui.ac.ir/index.php/him/article/view/47.

15. Chang R. Success through Teamwork: A Practical Guide to Interpersonal Team Dynamics (High-Performance Team Series) (Paperback). Homa-ye-Salamat 2006;2:63-5.[in Persian].

16. Ajami S, Fatahi M. The role of earthquake information management systems (EIMSs) in reducing destruction: A comparative study of Japan, Turkey and Iran. Disaster Prev Manag 2009; 18(2):150-61.

17. Ajami S. A comparative study on Earthquake Information Management Systems (EIMS) in India, Afghanistan and Iran. Journal of Education and Health Promotion 2012;1:27.

18. Society for Research & Initiatives for Sustainable Technologies and Institutions Organization (SRISTI). Disaster Management Information System' [online]. Available from: http://www.sristi.org/dmis/dmi_system. (accessed in 2009).

19. Afghanistan Information Management Service (AIMS) Project [United Nations Development Program (UNDP)]. Information Management for Natural Disasters: Pilot Project for Kabul &

Kunduz Province. 2005, Kabul and Kunduz Province. Afghanistan. Available from: http:// www.aims.org.af/services/sectoral/d_m/ dmis_for_afg_a_p_p.pdf.(accessed in 2005).

20. International Strategy for Disaster Reduction.Hyogo Framework for Action 2005-2015: Building the Resilience of Nations and Communities to Disasters. Extract from the final report of the World Conference on Disaster Reduction, 18-22 January 2005, Kobe, Hyogo, Japan.

21. International Strategy for Disaster Reduction.Hyogo Declaration. In proceeding of Conference on Disaster Reduction in Japan on 18_22 January 2005, Extract from the final report of the World Conference on Disaster Reduction, 18-22 January 2005, Kobe, Hyogo, Japan.

22. Eraslan C, Alkis Z, Emem O, Helvac C, Batuk F, Gümüsay U. et al. System Design of Disaster Management Information System in Turkey as a Part of E-Goverment. Istanbul, Turkey: Department of Geodesy and Photogrammetry; 2004. http://www.cartesia.org/ geodoc/isprs2004/comm2/papers/139.pdf.

23. Dedeoglu N. Role of the Turkish news media in disaster preparedness. In Proceeding of International Disaster Reduction Conference (IDRC), From August 25 to August 29, 2008. Switzerland, Davos.

24. Doi K, Kato T. Real time Earthquake information system in Japan. American Geophysical Union, Fall Meeting; 2003. Abstract # S21B-03-12/2003.

25. Yaliner O. Description of Urban information system and emergency management concepts, examples in Turkey and in the World. In Partial Fulfillment of the Requirements for the Degree of Master of Science in the Department of Geodetic and Geographic Information Technologies a Thesis Submitted to the Graduate School of Natural and Applied Sciences of the Middle East Technical University, January 2002.

Chapter 5

Learning from Lisbon: Contemporary Cities in the Aftermath of Natural Disasters

Diane Brand[1] and Hugh Nicholson[1]

[1]Victoria University of Wellington, New Zealand

INTRODUCTION

Cities survive earthquakes and rebuild with improved urban strategies, architectural designs and building technologies. This chapter will present a detailed analysis of the response, recovery, planning and rebuilding processes after the Lisbon earthquake and compare them to the recent events in Christchurch, New Zealand. Archival research from libraries and museums in Portugal will inform the historical analysis, and interviews and official documents from New Zealand relief and reconstruction agencies will underpin the contemporary analysis. The study aims to identify opportunities and challenges in facilitating good urban design in the process of recovering from a natural disaster, using case studies which are separated by over 250 years, but which both attest to the centrality of urban design in the reconstruction process.

LISBON 1755

This section will look at the Lisbon earthquake and its aftermath, with a view to understanding how design, leadership and governance processes contributed to the production of an 18th century, state-of-the-art urban quarter in the wake of a national tragedy. Particular attention will be paid to the coincidence of enlightened political, economic and technical skills which were judiciously applied to the re-planning of the city. By many counts the thinking was modern, and bears a worthwhile comparison to the recent seismic events in Christchurch, which will be the subject of the second part of the chapter.

THE EVENT

An estimated 8.4 magnitude earthquake in Lisbon on the morning of All Saints' Day in 1755 reverberated throughout the Iberian Peninsula, Madeira, the Azores, and North Africa. Tsunamis affected the Caribbean and the Atlantic coasts of Europe [1]. Of a population of 250,000, up to 20,000 people perished in the quake, tsunami and fires that followed [2]:

At this moment the earth shook, the sea rose up foaming in the harbour and dashed to pieces the ships lying at anchor. The streets and squares were filled with whirling masses of flame and cinders. The houses collapsed the roofs crashing down on shattered foundations. Thirty thousand inhabitants were crushed beneath the ruins [3]

Aid came from Portugal's colonies and her allies and trading partners England, Germany and Holland. Strategically, local merchants donated a 4% surcharge on imports to the relief effort which gave them critical political influence in determining new land uses for the reconstruction [4].

Figure 1: Panoramas of Lisbon before the 1755 earthquake [various authors] Museu da Cidade de Lisboa.

RECOVERY: A VISION OF ECONOMIC AND POLITICAL REFORM

The man in charge of the reconstruction effort, the Minister of State (1756-1777), Sebastião José de Carvalho e Melo, was decisive and commanding in his response to the disaster, relief and the forward planning of the city and his leadership in the crisis cemented his political power over his adversaries until the death of his supporter King José I in 1777. The King would make him the Marquis of Pombal 15 years after the earthquake in recognition of his leadership during the crisis. In a swift and articulate response to the emergency, the city was immediately surveyed and new construction was prohibited. Looters were publicly hung and able-bodied deserters were prevented from leaving the city and pressed into relief work by the army. Monasteries and public squares were filled with the homeless, and tent cities occupied by merchants and nobles sprouted (King José I and his family occupied an extensive tent and pavilion court in the hills of Ajúda at the edge of Lisbon for some months after the earthquake). Pombal moved about the city directing the recovery operation from his mobile headquarters a carriage which he had commandeered from the royal family.

What followed in the next two years was the ruthless and all-encompassing implementation of a radical plan which would change the political landscape and dramatically improve Portugal's economic position in Europe. The Terreiro do Paço (Figure 1) or Palace Square, had evolved as an elongated, spatially contained, urban space at the edge of the Tagus River in association with the Ribeiro Palace. In the wake of the earthquake the space was reconfigured, along with the adjacent central Baixa district, as the Praça do Comércio, a formal axial square surrounded by monumental public edifices where the business of an empire could be effectively conducted.

Pombal's previous roles as political envoy to London and Vienna and Minister of Foreign Affairs and War (1750-1756), where he oversaw town planting in Brazil, had seen him develop sophisticated ideas on mercantilist economic reform and coherent town planning. In the reconstruction of Lisbon's heart he found the perfect vehicle for both. The aim was to create a modern political centre where commerce could thrive. He had, in his ministerial role, overseen the dismantling of the Portuguese inquisition, the secularisation of education, and the nationalisation of industry [5]. He therefore favoured an institutional shift away from the old nobility (whom he considered corrupt and impractical) and the Jesuits, to the city's commercial elites who had helped finance the reconstruction. A strategically timed and implemented legal re-configuration of property ownership in the Baixa, transferred land from the aristocracy and ecclesiastic authorities to emerging merchant elites, whose collective economic enterprise would eventually succeed in rebuilding Portugal's indigenous economy. The waterfront square was edged with public buildings, which encompassed business, city government, and customs and exchange, in an effort to stimulate local trade and industry, and reverse high local unemployment levels and the traditional dominance of foreign merchants.

Urban renewal in the aftermath of the earthquake was appropriate for other reasons. Portugal had entered an era of nation-building with the consolidation of her borders in in 13th and 14th centuries. By the 16th century, Lisbon's image and role as the capital of a nation and a vast empire stretching from Angola to Macau was considered important, and urban embellishment was funded by gold and diamonds from Brazil. During the reign of King João V significant new urban projects were planned in the west of the city to boost the capital's status. The Águas Livres Aquaduct in 1728 is the best example of a new project that combined the provision of infrastructure with urban-scaled monumentality. Other projects included the construction of the vast and extravagant convent -palace at Mafra, the interior embellishment of 65 medieval and baroque churches and the building of dozens of new places of worship in the neoclassical style.

More significantly, Enlightenment-thinking from Northern Europe opened up the possibility of a scientific explanation for disaster rather than a religious one, rendering reconstruction a rational rather than a superstitious exercise, while the destruction of a splendid European court generated political uncertainties and revolutionary possibilities. As part of the reconstruction exercise, Pombal surveyed the populace in search of technical and scientific data about the earthquake, and his findings pointed to new methods in construction. Liquefaction had been observed in the riverside areas, so new buildings there were constructed on timber piles driven into the soil to act as anti-seismic stabilisers. Timber buildings had survived the quake better than masonry buildings, so new buildings had internal seismic frames added to their fabric. The new streets were widened so that even if the buildings on both sides had collapsed, there would remain an evacuation passage between them [6].

Ana Araújo [7] suggests that the press of the time stimulated a pan-European debate about issues of pragmatic responsiveness to natural disasters and unity of action across national borders. It was certainly the first example of a truly international relief effort after a calamity of such magnitude. Fact and fiction merged in the minds of the populace, fuelled by exaggerated emotive accounts of the catastrophe printed in cheap popular news pamphlets [8]. These served to fuel superstitious terrors, displays of religious fanaticism and dire predictions associated with natural phenomena such as Halley's Comet which was expected to appear in 1757 or 1758. The auto da fé, (act of faith, or public burning of heretics) held in the Terreiro in June 1756 was devised as a collective atonement for the earthquake of the previous year [9].

The event also presented political opportunity for the Minister of State to distribute anti-Jesuit propaganda in the capitals of Europe via anonymously authored opinion pieces in major newspapers [10]. Pombal co-opted some of the most influential men of the day in literature and science [11] to disseminate news and views about earthquake recovery, the necessity of repressive civil protection measures adopted post-disaster, and public health and welfare interventions. In the interests of civil obedience, state terror replaced the religious fear of the past.

EXPERIENCE AND EXPERTISE

The group charged with the recovery and rebuilding of Lisbon were an elite cohort of military engineers whose routine duties involved cartography, architecture and town planning, including the design and construction of infrastructure, (roads, aqueducts, ports, defence structures and fortifications) and associated buildings. They were a geographically mobile and flexible group of men, deployed in situations ranging from the frontier towns at outer reaches of the global Portuguese empire, to its busy cosmopolitan centre of trade and commerce. Their expertise embodied a high level of practical knowledge gained from work in the field as well as state of the art scientific knowledge. Their training via an apprenticeship system at the University Coimbra, represented a productive intersection of the knowledge embodied in foreign treatises such as those of Alberti and Vitruvius, and the real world they encountered in their active service.

Their approach to the built environment appears to privilege the site-specific adaptation of useful typologies, while giving a high priority to public space and infrastructure (especially port infrastructure, given that so many of the Portuguese colonial settlements and *feitórias* (trading posts) were coastal)[12]. Their sensitivity to the scale differential between the city and the building via a unified architectural language is a notable aspect of the Lisbon plan. Many of the best military personnel were in Lisbon at the time of the earthquake, and others joined them for the special mission of rebuilding the capital. Their designs were infused with the utopianism of the Portuguese school of the rational and civic-minded town planning demonstrated in new world cities such as Goa, Rio, Macapà and Luanda [13].

In 1910, cavalry officer Christovam Sepúlveda published Manuel da Maya e os Engenheiros Militares Portuguêses no Terromoto de 1755 (Manuel da Maya and the Portuguese Military Engineers in the 1755 Earthquake) [14] which identified Maia's central role as the strategist in the planning and rebuilding exercise. Jacôme Ratton [15] identifies military architect Eugénio dos Santos as the person who formulated the principles of the reconstruction.

Sepúlveda highlights two important aspects of the organisational culture of the military engineers. The first was their adherence to hierarchy, discipline and teamwork in military operations and the second was the heavy investment they made in strategic planning. Maia's role in executing both these practices is evident in the terms of reference for the reconstruction that were which embodied in the Dissertacão and other critical supporting official documents published in the late 1750s. Manuel da Maia's 1755 Dissertacão [16] describes the strategic solution for the rebuilding of Lisbon after the 1755 disaster. In addition to emphasising leadership, Sepúlveda deemed the teamwork, training and field experience of this group of elite military engineers in the colonial realm as essential preparation for the task.

There was ample regional and local precedent for disaster recovery as earthquakes had plagued Lisbon since its founding. A seismic event on the 26th of January 1531 had struck the city with equal force, but produced a far less acute political reaction due to the relative social stability at the time [17]. The Portuguese had certainly benefited from the study of other disasters such as the 1666 fire of London and the numerous volcanic events of in Sicily, particularly those affecting the city of Catania. Maia had been involved for many years in surveying and engineering projects around Lisbon, and this formed the basis of his knowledge to reconstruct the city. Critically he had been involved in the surveying and building of the Main Fortification Line in 1716, the Águas Livres Aquaduct, between 1720 and 1730, and the Santa Isabel survey in the 1740s. This deep knowledge of the terrain subsequently allowed him to swiftly prepare the alternative plans. He authored the Dissertacão which described the methods and processes leading up to the publication of the 1756 plan (Figure 2).

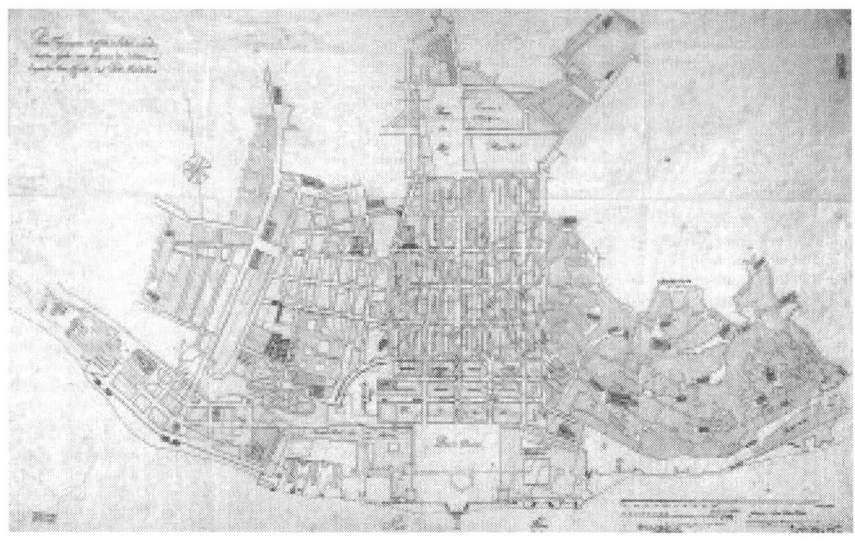

Figure 2: A Topographical Plan of the city of Lisbon [Carvalho and Mardel] 1755 Museu da Cidade de Lisboa.

PLANNING: THE DISSERTACÃO

The town planning principles for rebuilding Lisbon were established within 5 months of the event, in Part I of the Dissertacão, which directed that options for both rebuilding and relocating the city be considered. There was an impetus to rebuild and relocate the court in a better place (either at Bélem or São João dos Bem-Casados) which opened up the possibility for locating new public buildings in the Baixa. Layouts were subsequently developed by military architects Eugénio dos Santos and Carlos Mardel.

Part II of the Dissertacão was comprised of a legal decree in May 1756 that governed land ownership, construction and finance for the rebuilding of the devastated centre, and a set of development rules for areas of expansion at the periphery. Part III of the Dissertacão was comprised of the alternative plans, which covered a range of rebuilding and relocation options. Maia presented six options investigating variations of street widening, ground re-levelling, adjusted building heights, all contributed to a radical new block morphology, residential-building typology (with trade and craft premises at ground level) and sense of urban scale [18]. A prefabricated, technically innovative and stylistically simplified four-storey building type was developed. This building ingeniously incorporated fire walls that extended above the roof line and earthquake resistant diagonally braced, timber load-bearing cages called gaiola that sat independently within the masonry perimeter walls. Building components were prefabricated on a large scale off-site, allowing speedy on-site assembly with minimal traditional craft involvement [19].

The plan ultimately chosen, incorporated the following mechanisms and characteristics which are documented by Claudio Montiero [20]:

- A standardised building type
- A standardised construction solution incorporating fire separations and seismic frames
- A rational and generous public space network based on pre-existing places
- The use of the rubble from the ruins to raise and regularise the levels of the Baixa by 1.2 metres, and reclaim new land from the Tagus
- A disregard of property ownership boundaries (especially the location of churches) so as not to compromise the rigorous geometry of the new plan

This formed the basis for the legal provisions of the legal decree of June 1758 which finalised the plan. It had taken two years for the plan and the legal and financial exchanges to align. A year later, in June 1759, re-building finally commenced.

These critical ownership reforms relied on Pombal's public credibility and private influence at Court. The period was dogged by multiple conspiracies, including a palace coup in 1756, and an assassination attempt on the King in 1758. These behind-the-scenes machinations lengthened the plan's implementation time, but unlocked the heart of the city as a centre of trade and commerce, thereby better serving the emerging merchant class and challenging the historic power of nobility and clergy. The plan largely preserved the places and names of the historic city and retained the location, hierarchy and functions of three main squares (with the re-naming occurring later). Churches that had been free-standing were now integrated into the Baixa blocks.

The Praça do Comércio doubled the size of the former Terreiro do Paço, by reclaiming land from the river Tagus. The new square's symmetry, focusing on an equestrian statue (flanked by the animals of Portugal's far-flung continental empires), and the triumphal arch to the main street, Rua Augusta, constituted the axis of the plan as a whole. The post-earthquake square had a statue of King José I at its heart but perimeter uses were designated for functions of state, with the palace itself relocated to the city edge at Ajúda [21]. While symbolically, a royal statue still stood at the centre of Lisbon, functionally it was now a place for commercial enterprise (Figure 3). Blocks were configured in a simple proportional and compositional system that supported elegant and environmentally comfortable street sections and public-space footprints. There was a hierarchy of three main streets, each 60 palms wide, that were named for the guilds (Rua da Prata, Rua do Ouro or Silver and Gold Streets respectively), cross streets that were named for church and parish interests, and North-South streets, each of widths of 40 palms, that were also named for guilds.

Figure 3: Praça do Comércio [Rick Allender] 2007.

The streets determined the building fabric. The building heights were to be no higher than the width of the streets and thus set a new standard for access to light and air in an urban building. A system of dividing dwellings by floors, with retail at ground-level and artisans' workshops at basement-level, was developed. The arrangement of three floors of identical apartments plus an attic above the ground-level floor was the origin of modern mixed-family ownership in Portugal, and the typology represented the potential for a mixing of classes in one edifice, each with its own separate entrance. Both shops and apartments were built to be rented, thus facilitating urban housing as a form of property investment.

The plan's implementation was directed by José Monteiro de Carvalho, who materialised the abstract rules for scale and the architectural features of the urban blocks at a large scale, while enforcing technical standards such as fire compartments between tenancies, new sewer locations and the cage structures. He was also attentive to the finer details of serial design elements required for cast-steel balconies, the ashlar masonry trim to building bases and openings, and consistent window and door joinery profiles (Figure 4). Up to this time he had been in charge of demolition, which earned him the nickname Bota-Abaixo or knock-it-down.

Figure 4: century standardised building types in Chiado District [Diane Brand] 2004.

This aesthetic and technical system became the basis for re-planning Lisbon as a whole, especially around the new palaces, and in areas developed at higher densities in the western part of the city. The scheme set the tone for a future direction of urbanism in Lisbon that embodied a new rational Cartesian pattern but which was firmly anchored in a traditional morphology. Over time the plan set aesthetic, technical and legal precedents but these were not fully appreciated until the modern era, when separate built environment professional disciplines such as architecture and planning emerged and their histories were fully researched.

REBUILDING

Claudio Monteiro [22] suggests that legislative reform enacted during the reconstruction of Lisbon was driven by the plan's necessary transformation of the structure of urban property ownership, and the careful reconciling of individual rights with the security of future investment. The reconstruction plan and the resultant legislative reform were the tools that brought about eventual political and economic reform. Pombal's aim was to consolidate the power of the King while at the same time modernising the nation's legal, economic and social structures.

The measures used to achieve these included:

- The surveying of existing buildings at the time of the earthquake to avoid disputes during reconstruction, especially when an overhaul of the land ownership arrangements was contemplated (Wren's plan for London after the fire in 1666 had been frustrated by an inability to rationalise the nobility's ownership of the large estates in central London).

- A prohibition on constructing or reconstructing buildings before the plan's approval:

- Outside of Lisbon, to stop the city growing randomly and

- Within Lisbon, to prevent the rebuilding of buildings partially destroyed by fire (less than one third of the original buildings were in a habitable condition and no alternative accommodation existed apart from tent cities and timber shacks erected in public spaces).

The plan was approved two and a half years after the earthquake, and the first lots were reconstructed three and a half years later. Nevertheless, illegal urban development had sprung up in spite of harsh enforcement of the decrees.

- Freezing rents and freezing the price of construction materials, to combat speculation, eviction and exploitation around the shortage of construction materials and rental accommodation. This was done by restricting any new lease agreements to perpetual leases or long-term rental contracts.
- Creating the conditions for legal, religious and political reform by freely compensating and transferring pre-existing property ownerships into newly agreed formats
- The complete demolition of the Baixa to make way for a despotic but utopian and progressive plan [23]

COMPENSATION

Land within the Baixa was immediately appropriated by the state and re-allocated, with preference given to existing land owners, leaseholders or administrators for nobles, the church or the crown. Compensation was based *only* on site area, and not the post-earthquake building condition. New lots were allocated on the condition that redevelopment would be completed within five years, effectively rendering the exercise a land re-adjustment operation rather than an exercise in eminent domain, while preventing long-term speculation of development leap-frogging.

Undersized lots, oversized lots and lots eliminated by the creation of new public spaces or streets were paid fair land swap or cash compensation in proportion to the frontage width of the site. Cash payments were necessary since there was a greater total area for public space in the new plan and therefore an undersupply of new sites. Maia proposed a proportional reduction of all buildable areas to account for improved amenity as a result of more public space in new areas. The chief surveyor of the inspections, Alexandre José Montanha divided the Baixa into seven zones of value, thus setting up a financial mechanism by which properties were exchanged or compensation calculated with a 'premium payment' embedded for superior sites adjacent to public space.

In this way the plan created value. The overall effect was to replace certain types of landowners (nobles and secular clergy) with merchants, sparking what Subtil called 'political earthquakes' in Portuguese society [24]. The compensation system and plan stimulated investment from the business community who had financed the reconstruction (via credit or purchase). This in effect led to a significant redistribution of wealth, a consolidation of economic power among the middle class, and a new degree of upward social mobility. The move also unlocked the encumbrances and liens strictures embedded in the medieval property codes that had inhibited clean development processes within the city.

Complete execution of the plan took over 40 years. An initial displacement of Lisbon's population to the west immediately after the earthquake inhibited the uptake of property in the centre. The ancient elites also retreated, taking the court sector with them.

AUTHORITARIAN PROCESSES

Authoritarian processes were the key to the effective reconstruction of Lisbon. Pombal was appointed by the King to his position as Minister of State and he used this mandate to centralise political power by removing the Senate from the state decision-making processes and from the implementation of the Baixa plan, thereby breaking a longstanding tradition of local autonomy in planning and taxation matters. The institutional makeup of the reconstruction process evolved as the pragmatics of the situation dictated, with two complementary bodies emerging: The first was the Lisbon Neighbourhoods Inspectorate as the civil defence responder in 12 neighbourhoods city-wide. This agency was also instrumental in clearing debris, removing and burying bodies, executing surveys and re-allocating land. The second was the Public Works Department which was formed to implement the plan, and projects including public spaces and new buildings. The technical team, comprising of army officers in the civil administrative hierarchy, originated at the Lisbon Public Works Draughting Office (*Casa do Risco da Obras Públicas*) which later became the Public Works Department. This reduced, focused and disciplined chain-of-command, facilitated the absolute control required for such sweeping changes to the urban space configuration and the resulting shift in political and economic hegemony.

The first order of business was the creation of the public realm, with new streets, squares and gardens. Public health was foregrounded with upgraded sanitary infrastructure, water supply and transport systems given priority. The construction of essential public buildings for trade and business continuity such as the Customs Building in Praça do Comércio was also critical as they served an influential special-interest group. Reconstruction took roughly until 1807 to complete, ironically coinciding with the royal family's flight to Brazil as Napoleon's troops massed on the border during the Peninsular Wars.

URBAN DESIGN OPPORTUNITIES

A pamphlet at the time of the 1755 disaster optimistically stated that 'Lisbon could not have suffered a more fortunate tragedy', indicating the potential the populace saw in the reconstruction process [25]. A major aspect of this fortune clustered around the implementation of a good example of urban design, one that was ground-breaking for its time and which still ranks as outstanding. The plan proceeded with a new gridded layout for the Baixa quarter, based on the disaster and re-planning precedents of the fires in London and Rennes, and the earthquake in Catania and planning precedents such as the 1620, 1673 and 1714 extensions to Turin [26]. Among precedents for the Baixa Plan were Wren's 1666 plan for London (new street alignments and property subdivision), new Turin (the regularised geometry of public space and city blocks) and the Place Royale, Place Vendôme and the Royal Palace at Bordeaux (the continuous articulated facades, marked entry points and arcaded bases of the buildings which framed the public space) [27]. It represented a successful example of contemporary urban disaster and urban design knowledge of in 18th century Europe.

The Terreiro do Paço was reconstructed and renamed the Praça do Comércio (Figure 5). Rubble from the earthquake was recycled (eliminating the disposal problem) and an area of land equal in size to the original square was reclaimed, extending the urban platform into the Tagus. The reconstruction of Lisbon presented an opportunity to integrate the waterfront square into the urban fabric. The new monumentally scaled square used symmetry and architecture to integrate a complex of buildings embracing the space into the urban fold, and created a powerful central axis penetrating into the city behind via Rua Augusta, thereby linking the square to the Rossio (Lisbon's other principal square) beyond. Pombal's project redressed the problems that had beset the Terreiro as an urban square. Certainly the architectural and urban legibility of the square was enhanced, the buildings were better scaled to city blocks and there was more permeability to the Baixa, with vistas along Rua da Prata, Rua Augusta and Rua do Ouro. However, the authoritarian method of delivery required the subordination of individual property rights to the public interest, with new development and construction precisely defined within the strict constraints of the plan. The effect was also to subordinate architecture to urban design, with exacting and specific controls placed on building envelopes, construction methods, uses, appearance and materials.

Figure 5: Praça do Comércio [Susana Pereira] 2011.

The following section discusses urban design initiatives arising from a series of devastating earthquakes in the new world city of Christchurch, New Zealand and compares them to the 18th century event in Portugal. Two years after the earthquakes, urban design strategies to rebuild the broken city have been formulated but are not yet implemented.

CHRISTCHURCH 2010 AND 2011

Christchurch was planned in the mid-19th century as the last Wakefield settlement to be established by a private colonization enterprise called the New Zealand Company. The company was involved in the establishment of six urban centres in New Zealand: Wellington [1839] Wanganui [1840], Nelson [1841], New Plymouth [1842], Dunedin [1848] and Christchurch [1850]. The New Zealand Company brought more than 9,000 hopeful settlers to New Zealand up until 1843 [28], with each of the towns achieving numbers of between 1000 and 4000 in the first years of settlement. Christchurch was sponsored by the Church of England and aspired to recreate a stable agrarian, hierarchical society on fertile land, between the Pacific coast and the Southern Alps on terrain purchased from the local Ngai Tahu tribe. A rectangular grid was surveyed onto flat swampy land and adjusted where necessary to accommodate the Avon/Otakaro River, which meandered across the site. The diagram of the city incorporated the grid, the river, two diagonal roads registering the principal transport routes to the port of Lyttelton and the main road north, a cross-shaped central square (where a cathedral was later built), a market adjacent to the Avon River, two asymmetrical peripheral squares, and parklands providing generous and varied public and recreational space. In June 2010 the city was New Zealand's second largest, with a population of 376,700.

EVENTS

In the early morning of Sunday the 4th of September 2010 a magnitude 7.1 earthquake located 40 km west of Christchurch on an east-west fault, not previously identified, struck the city. The fortuitous timing of the event explains the lack of fatalities but there was widespread damage to unreinforced masonry (URM) structures [walls and chimneys] in Christchurch and surrounding towns. Nineteenth century URM shop fronts collapsed into main thoroughfares in the CBD, and there was widespread liquefaction to eastern suburbs' residential areas close to rivers. This earthquake triggered a series of aftershocks that moved progressively closer to the city, culminating in a shallow, 5km deep, M 6.3 earthquake on February 22, 2011 (Figure 6) centred at Lyttelton. This too occurred on an unidentified fault. This second major event resulted in 181 deaths, with more than half of these occurring in the collapse of the Canterbury Television Building. This earthquake caused further significant damage and collapse to URM structures in the CBD, more liquefaction to eastern suburbs' residential areas close to rivers and rock falls and landslides to the south and southeast. The result is that between 50% and 70% of buildings in the CBD are likely to be condemned for demolition with the discovery that many structures are out of vertical axis alignment due to differential settlement and the realization that the repair of others is uneconomic.

Figure 6: A 6.3 magnitude earthquake strikes Christchurch on February 22nd 2011 [Gillian Needham].

The civil defence response was immediate, with international urban search and rescue terms arriving from nations such as Australia and Japan to relieve local emergency services. Emergency response centres were swiftly established in parks, and sports and community buildings, with temporary accommodation provision in the form of tents, caravans, and prefabricated houses deployed by the Department of Housing and Construction. The entire CBD was cordoned off and placed under police and army jurisdiction for an extended period, leaving many businesses without access to their premises. The slow and uneven process of insurance compensation has led to businesses relocating to other centres in New Zealand or locally to the western edge of the city close to the airport. The entertainment centre of the city has re-established itself in the west along Riccarton Road. With more than half of Christchurch's listed heritage buildings (250) located in the CBD, the city's patrimony has been particularly hard hit with more than 100 demolitions to date. The iconic Anglican Cathedral, the Catholic Cathedral of the Blessed Sacrament and the Canterbury Provincial buildings suffered significant damage (Figure 7).

Figure 7: Canterbury provincial buildings post-earthquake [Diane Brand] 2011.

RECOVERY

After the second earthquake it was clear that existing central and local government agencies were not equipped to facilitate recovery operations, and The Canterbury Earthquake Recovery Authority (CERA) was formed on March 29, 2011 as a special government agency for the co-ordination of the recovery and rebuilding activities in Canterbury. The Canterbury Earthquake Recovery Act [29] gave unparalleled (in

New Zealand terms) authoritarian powers to the Minister of Earthquake Recovery, The Right Honourable Gerry Brownlee, although in practice the powers have been exercised only with the agreement of the cabinet (the government's executive level ministerial group). In particular the CER Act allows a recovery plan approved by the minister to override the requirements of New Zealand planning legislation frameworks embodied in the Resource Management Act, the Conservation and Reserves Acts and large parts of the Local Government Act (although not the funding provisions, and the Land Transport Act).

The government's response to the Canterbury earthquakes occurred in an environment of mistrust between national and local government, characterised by the dissolution of the regional council, Environment Canterbury, in March 2010, and the quite different leadership styles of the mayor and the minister. Part of the reason for the distrust lay in the different underlying political philosophies, with the ruling national government espousing a 'shrinking government' position together with the sales of government assets as a means of reducing the national deficit. In contrast, local government in general, and in particular the Christchurch City Council (CCC) supported maintaining the level of local government services as a minimum, with the CCC pursuing a clear position of holding onto city assets in city-owned holding companies and using these to generate income or reduce tax liabilities.

In 2011, the government introduced an amendment to the Local Government Act aimed specifically at limiting the services local governments could provide and the levels of rate increases they could introduce. The CER Act specifically excluded the minister from making changes to the funding provisions of the Local Government Act and there have been continuing discussions about the allocation of costs between national and local government ranging from the emergency response costs to repair and rebuilding costs. The national government has clearly stated on a number of occasions that it believes that the CCC should sell some of its assets to fund the recovery bill.

The CER Act established a new government agency to oversee the recovery of Canterbury and the government's investment in the rebuilding of Christchurch. The act specifies that the minister can direct the city council, but does not clearly establish the respective roles of the organisations, or create any direct organisational links or lines of management apart from general requirements to consult.

The newness of the government agency CERA, coupled with the pre-existing responsibilities of the council, has led to a lack of clarity about their respective roles, with duplication happening at a number of levels between the two organisations.

The CER Act specifically required the CCC to develop a draft recovery plan for the central city in nine months for the minister's approval, including public consultation. Planning for the rest of the Christchurch metropolitan area was the responsibility of CERA. The Draft Central City Plan (CCP)[30] was completed in eight months, however the minister spent a further seven months reviewing it. When the minister received the draft he endorsed the vision contained in the first volume, with the exception of the proposed transport changes, but he set aside the proposed regulations for further investigation [31]. The 'blueprint' plan [32], subsequently approved by the minister, broadly adopted the range of major infrastructure projects proposed in the Draft CCP and retained the majority of the proposed regulations including the reduced height limits. The major changes from the Draft CCP was the removal of the regulations requiring improved environmental performance from buildings (the BASE assessment developed with the NZ Green Building Council), the removal of the financial incentives for rebuilding proposed by the council and the removal of the majority of the transport provisions pending further investigation.

The Canterbury earthquakes resulted in extensive land damage and areas of liquefaction, particularly in the eastern part of the city (Figure 8). The resulting changes in elevation included some areas in the Port Hills rising by up to 500 millimetres while areas around the estuary and Avon River subsided by more than 500 millimetres (Figure 9). In extensive areas the cost of land remediation, flood protection and/or the restoration of services made it uneconomic to rebuild on the terrain. The government assessed all residential land in Christchurch and the surrounding towns based on extensive geotechnical studies and eventually classified them as either 'green' (fit to rebuild) or 'red' (unfit to rebuild). Subsequently, the government has set about purchasing more than 6,000 houses in the residential red zone based on the 2009 rateable valuations. The houses are generally either clustered in low lying areas around the Avon River and Estuary or vulnerable to rock fall in the Port Hills.

Figure 8: Liquefaction in the eastern suburbs of Christchurch (2011) New Zealand Aerial Mapping Ltd for LINZ.

The retreat of settlement along the Avon River and Estuary in Christchurch has provided a microcosm of the kinds of issues likely to be faced by many coastal cities worldwide, if sea level rises predicted over the next century occur [33]. The model of strategic retreat, from vulnerable areas may become relevant in many other areas. While the Christchurch model has addressed the issues of strategic retreat and attempted to manage the economic impact on residents, no attempt has been made to address the impacts at a community level. However, a map showing the areas where 'red zoners' have relocated reveals a scattered pattern determined by the prices and availability of houses, rather than any managed attempt to relocate communities.

The CER Act facilitated the immediate use of earthquake rubble for reclamation work to extend the container port at Lyttelton. This would have been difficult and protracted under the RMA. The port is one of the key economic drivers for the Canterbury economy, and the port extension reflects the changing scale and technologies of port logisitics. The CER Act has also been used to fast-track residential subdivisions, thereby short circuiting the currently protracted consent and environment court processes. The intention has been to free up residential land so that people who have been displaced by the earthquakes, and workers arriving in Christchurch to assist with the rebuild can be adequately housed. In doing so, the minister has confirmed the overall urban form proposed in the Greater Christchurch Urban Development Strategy which sets out urban limits, greenfield residential areas and housing densities, targets for intensification, urban design outcomes and key transport corridors (although these measures are currently being challenged through the courts). The need to dispose of a huge quantity of rubble in a very short timeframe, and to expedite the provision of new housing stock to replace that damaged and destroyed are common issues for cities struck by earthquakes. Both Lisbon and Christchurch used authoritarian powers to address these issues in a timely and economically beneficial manner.

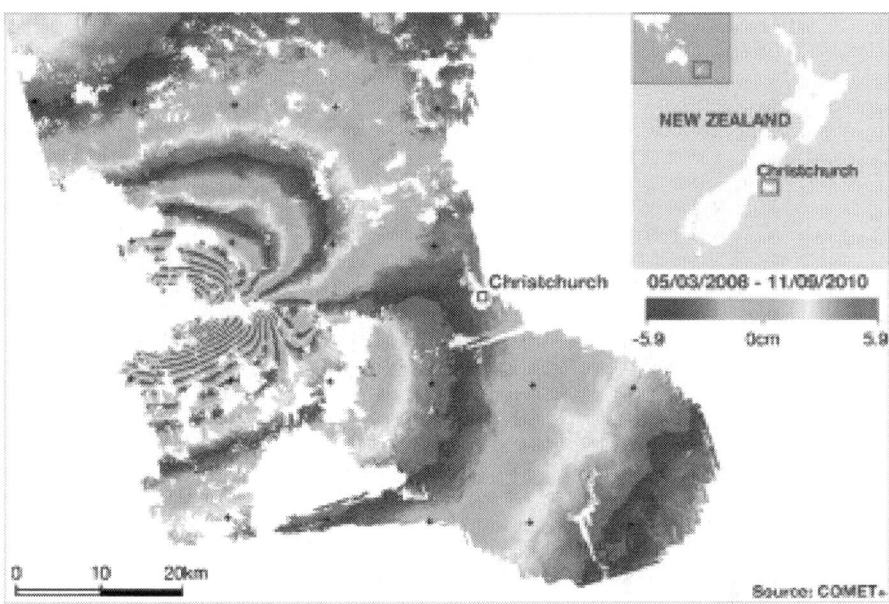

Figure 9: Lidar Map of Christchurch showing ground level changes (2010) COMET.

COMPENSATION

After a wave of earthquakes that plagued New Zealand between 1929 and 1942 (the worst of which was the M7.8 Hawkes Bay earthquake of 1931 which completely devastated the cities of Napier and Hastings), the Labour Government created the Earthquake and War Damage Commission. The commission eventually insured all residential properties, including the land (up to $100,000 for buildings and $20,000 for contents) against damage from natural disasters [34]. Commercial property must be privately insured. In early 2010, the EQC had reserves of NZD 5.6 billion backed by a government guarantee. However the total insurance estimate for the Christchurch earthquakes is upwards of NZD 30 billion [35] and represents the worst natural disaster in a developed nation relative to size of economy. Private insurance fared no better, with the market dominated by only 5 principal providers, several of whom have struggled to pay out claims without government assistance. The compensation for damage is an on-going saga of bureaucratic complexity, delays and individual suffering which has been exhaustively documented by the nation's media. Policy neglect had contributed to the underinsurance of the majority of New Zealand homes, with EQC covering a quarter of the average value of a New Zealand home, and maximum premiums still at their 1973 level of NZD 67.50 per annum in spite of advice to increase these in 2008. A premium for land insurance had never been charged, but this type of damage turned out to be the most expensive. While the annual EQC premiums were raised in the immediate aftermath of the earthquakes, policy revision in the insurance sector is still under consideration. Discussions include the introduction of land insurance, better processes of alignment between land use decision making in territorial local authorities and disaster insurance agencies, and the level of future risk-sharing between EQC and private insurers [36].

EXPERIENCE AND EXPERTISE

Local government manager Warwick Isaac, who was overseeing the demolition of buildings in Christchurch, was appointed to lead the Central Christchurch Development Unit (CCDU) which is tasked with leading the rebuilding of the central city. Dubbed 'the demolition man', a Tom Scott cartoon has him lamenting "If I had known you were going to put me in charge of the rebuild Minister I wouldn't have pulled so much down." Both the council's Draft Central City Plan and the 100-day Blueprint subsequently produced by the CCDU were prepared by multidisciplinary teams led by urban designers. Although the approaches were fundamentally different, with a community-led bottom-up process for the Draft Central City Plan and a technocratic top down 'masterplan' in the Blueprint, both documents were 'design-led' in that they used design as the key method of developing the plans with built environment professionals (urban designers, landscape architects, architects) holding the key leadership roles responsible for developing plan content.

The Draft Central City Plan was developed by a multidisciplinary in-house team of council staff that included seconded team members from national and international consultancies including Gehl Architects from Copenhagen. Hugh Nicholson, the Principal Urban Designer at the Christchurch City Council, was responsible for delivering the content of the plan. The team included urban designers, architects, landscape architects, engineers, economists, planners, community advisers, communication specialists, sustainability advisers and recreation and open space planners.

The 100-day Blueprint was prepared by a consortium of design companies led by Boffa Miskell, a local company specialising in landscape architecture, urban design and landscape planning. The consortium included local architects Warren and Mahoney and Sheppard and Rout as well as specialist convention centre and stadium designers. The team was led by landscape architects Don Miskell and Rachel De Lambert and urban designer Marc Bailey.

PLANNING

Within three months of the February earthquake, an extensive public consultation exercise was undertaken with the people of Christchurch to help shape the future plan of the devastated city. The website shareanidea.org.nz generated 58,000 hits and engaged the public in four key areas: move (transportation), market (business), space (public place and recreation) and life (mixed uses), across traditional and social media networks. This was followed by an interactive expo, 10 community workshops, 100 stakeholder meetings and a professional competition for 5 selected sites. An unprecedented level of public participation generated 106,000 ideas over six weeks and these informed the development of the Draft Central City Plan.

One of the firm assumptions underlying the Draft Central City Plan was the maintenance of the existing street and land ownership patterns. In part this recognised the strong urban form provided by the existing grid and its heritage values. In part it was in response to initial estimates that suggested 50% of the buildings in the commercial core might be demolished (subsequently this looks to be greater than 60%). This implied that there was still a substantial residual value in the remaining buildings and services which the city could not afford to lose. In terms of broad urban design objectives, both the draft CCP and the Blueprint set out to provide an enhanced network of green open spaces based around the Papa o Otakaro (Avon River Park) and Cathedral Square, to rebuild a more compact intensive low-rise commercial core, to increase the number and density of inner-city residents, and to promote mixed-use developments in areas surrounding the core. The plan also proposed more sustainable transportation systems, including a light-rail system from the university to the CBD, that would eventually connect into a regional rail network and a grid of cycle-ways. The redevelopment clustered around a set of core projects seeded by government, which were designed to attract investment and rebuilding in the CBD and these included a greening of Cathedral Square, a sports hub, a convention centre, a central library and a hospital campus.

One of the more controversial urban design proposals in the Draft CCP was a reduction in height limits to 28 metres or seven stories. Christchurch was the first major Australasian city to propose a low-rise urban form, moving away from the modernist podium and tower model of development. The reasons for this were partly the high level of community support for low-rise buildings and their desire to create a more human-scale environment with better environmental conditions, including improved sunlight access and reduced wind funnelling. Additionally, economic modelling indicated that due to the increased foundation and structural costs required to build higher than six to seven storeys, the most economically viable built-form with the highest rate of financial return was in this height range. The final reason was to address the oversupply of commercial land in the core, by rebuilding a more consistent intensity of development over the area of the core, avoiding the spikes of oversupply and undersupply provided by the tower model. There was a strong backlash from the business community against the proposed height limits and this was one of the provisions that the national government set aside when it reviewed the Draft CCP. However at the end of the review, they reconfirmed the height limits based primarily on the economic impact assessment and the land supply issues.

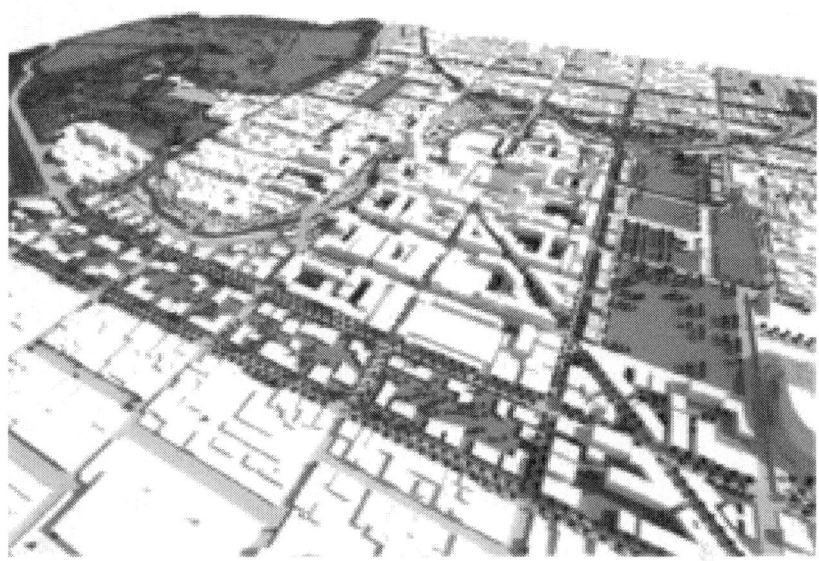

Figure 10: The CCUP Blueprint (2012) CCUP.

The Draft CCP proposed using built-form restrictions to further promote a compact commercial core, with a higher intensity of development through the use of incentives and by limiting development potential outside the core. The Blueprint adopted a far more interventionist approach by establishing a 'green frame' (reinstating nineteenth century parklands) and compulsorily acquiring large areas of land surrounding the core. The long-term future of the proposed frame is not entirely clear. Some parts of it appear to be intended as permanent open spaces, some parts such as the health precinct are earmarked for campus-style commercial development, while other parts appear to form a potential land bank for release once land in the core has been fully developed.

In summary, the Draft CCP adopted a multifaceted approach to recovery that incorporated a wider range of projects and implementation tools. The vision balanced incentives and regulation to deliver major catalyst and public space projects, alongside sustainability, housing, arts and transport projects. The blueprint focuses more deliberately on national government priorities, providing a regulated vision embodied in a range of catalyst projects that involve rebuilding critical public and economic infrastructure such as the hospital and the convention centre. At the second anniversary of the first earthquake, the city has started to rebuild with 1000 building consents in the past 12 months. Processes are being put in place by CCUP to fast track significant projects through an urban design board process while the CCC Urban Design Panel is doubling in size to cater for the anticipated increasing volume of resource consent applications.

CONCLUSIONS

Events in Lisbon and Christchurch stand apart in chronology, severity and extent, recovery management, and design outcome, but have sufficient in common to draw some interesting and relevant conclusions to 21st century disaster-response strategies. Prime amongst these is the use of urban design as a revitalisation mechanism, as it is a natural aspiration to want to rebuild a devastated metropolis anew, correcting the mistakes of the past by implementing new and state-of-the-art practices to envision a better city.

Lisbon suffered not only a cataclysmic earthquake but also a devastating tsunami and fire. Fortunately for Christchurch the latter two stressors were absent, and 250 years of improved planning, seismic and fire engineering performance, reduced the relative death toll from building collapse while generous provision of public space allowed the population to escape to safer areas. Christchurch, in achieving this high level of technical preparedness, is much indebted to Lisbon which pioneered many contemporary post-earthquake response strategies. Two hundred and fifty years before the terms resilience or sustainability entered the built environment lexicon, their guiding principles were applied in Lisbon. Pombal's engineers made the decision to rebuild in the same location, but not before investigating six alternative sites and researching the technical failures that led to the high death toll. In so doing they were embedding in the plan the future sustainability of the city and built in resilience for future seismic events, not only for themselves but also for others who chose to follow their example.

The civil defence emergency response was, for the first time in history, an international one, with Portugal's trading partners stepping in to assist. The necessary revenue for recovery was raised via import taxes levied by local businesses. Exploitative behaviours were curtailed by punishing looters, freezing rents and the price of materials. The immediate surveying and demolition of the area reduced the territory to a uniform and indisputable condition in terms of future claims. Prior to any rebuilding, a post-disaster analysis was conducted to establish which buildings had survived and why. These investigations led to a number of technical innovations that required a new formal and technical building typology. This simple and elegant solution to rebuild the city relied on three crucial pillars: the complete demolition of the devastated Baixa, the re-drawing of property lines, and an urban design plan that integrated these technical provisions into a best-practice vision (based on solid international precedent) for a commercial rather than an institutional centre for Lisbon. The implementation of these strategies was only possible due to the authoritarian nature of the governance system at the time, and an emergency response which delivered this power unilaterally into the hands of the Minister of State.

A similar approach is clearly not possible or appropriate in a modern democracy, although the potential exists and was contemplated within the New Zealand legislation introduced to affect recovery plans in Christchurch. The national government has committed substantial leadership resources and legislative support to the recovery of Christchurch and the Canterbury region. Of particular note has been the strategic land use withdrawal from the residential red zone, the on-going demolition of dangerous buildings, and the proposed major infrastructure and facilities as catalyst projects. The authoritarian powers provided through the CER Act have enabled these initiatives to occur with the minimum of delay or inappropriate process, although some delays have occurred due to the lack of clarity about the respective roles of the council and CERA. The Draft CCP proposed an integrated plan for the central city, while the blueprint approved by the minister focuses on rebuilding the major infrastructure and facilities and leaves out much of the 'glue' - the smaller scale projects that hold the big moves together. The absence of urban residential typologies or social housing to accommodate earthquake victims from the list of prioritised projects, overlooks the capacity and necessity of embedding these in the plan as community or capital investment opportunities in the way they were in the Lisbon plan. The council currently intends to continue with a number of these smaller projects in tandem, so the end outcome may well be the same albeit encapsulated in two plans rather than one. The major omission of the majority of the transport provisions is the subject of a further study.

The Draft CCP and the approved blueprint have both been led by urban designers and shaped by urban design propositions, in particular a low-rise, more intensively developed city based on economic factors with high quality green and public spaces shaped for people. The extent to which they can replicate the success of the Lisbon reconstruction is at least in part subject to international economic forces and remains to be seen. Equally important is the clarity developed around future urban design controls. While these exist in the Draft CCP, the minister has side-stepped council involvement in consent processes and set up a new consent authority to oversee central-city consents, with one representative each from the council, CERA and Ngai Tahu, (with no articulated formal role for Christchurch City Urban Design Panel). This body has a mandated consent turnaround of 5 working days as compared to the usual 20- to 85-day timeframe (depending on levels of compliance).

The top-down process enacted in the Lisbon earthquake or more recently the Kyoto earthquake, and the bottom-up process followed after Hurricane Katrina, sit at either extreme of the continuum of possible response management strategies to natural disasters. A balance between these extremes is more feasible, the balance depending on the socio-cultural and economic context and the governance systems in place at the time of the crisis.

Communication Technologies have had a major impact on response capability in the intervening centuries between Lisbon and Christchurch. Tsunami early warning systems and international media networks give instant alerts of impending disasters allowing preparation or evacuation. Cell phone networks, satellite communications and GPS tracking and positioning technologies contribute to more effective search and rescue operations. Collective media and social network platforms pressure reconstruction authorities to deliver on their promises in a timely manner. The internet collapses the time required for widespread public consultation leading to more effective community buy-in into new urban proposals.

In Christchurch, the lack of alternative design proposals from official sources has been a response to the short timeframes imposed by the government. This is less concerning or necessary given the wide consultation undertaken to reach the plan outcome. Again in spite of the 250-year separation and with different professional actors, Lisbon and Christchurch had good levels of technical expertise available to generate an urban design-led reconstruction effort using current contemporary urban theory around sustainability and resilience planning in combination with deep local knowledge.

The technical planning and architectural detail is not yet present in the Christchurch plan and will be managed by planning consent processes that have not yet been well defined. Lisbon provides an excellent model for reinventing a modern local urban type (a mixed-use, low-rise, multi-tenancy, structurally sound and fire-protected building) and designing an urban block morphology that reflects a historical vernacular. This will be the fabric that weaves the plan framework and the demonstration projects into a real city.

New Zealand government agencies in charge of engineering, building and construction standards have not yet integrated the lessons from the 2010 and 2011 earthquakes into upgraded performance codes. As a nation located on a chain of islands on the Pacific 'ring of fire', uptake of resilience strategies like those in enlightenment Portugal must encompass flexible governance systems, high-level technical expertise in national, regional and urban planning sectors, and innovative architectural, communications, engineering and material technologies. This in combination with communities helping themselves is the best insurance against future calamity.

REFERENCES

1. Baptista, M., Miranda, J.,Miranda, L. and Mendes V. (1996) Rupture Extent of the 1755 Lisbon Earthquake Inferred from Numerical Modeling of Tsunami Data, *Physics and Chemistry of The Earth, 21/ 1-2*, pp.65-70.

2. Araújo, A. (1997) *A Morte em Lisboa. Atitudes e Representações, 1700-1830 (*Lisbon: Editorial Notícias).

3. Voltaire (1759) *Candid, or All for the Best* (London, J. Nourse).

4. Kostof, S. (1991) *The City Shaped: Uban Patterns and Meanings Through History* (London,Thames and Hudson).

5. Maxwell, K. (1995) *Pombal, paradox of the Enlightenment* (Cambridge, Cambridge University Press).

6. Hamblyn, R. (2009) *Terra, tales of the earth: four events that changed the world* (London: Picador).

7. Araújo, A. (2006) 'The Lisbon Earthquake of 1755-Public distress and political propaganda', *e-JPH 4/1 pp10*.

8. Esperjo, C. (2005) 'Spanish news pamphlets on the 1755 earthquake: trade strategies of the printers of Seville' in T. Braun and J. Radner (eds) *The Lisbon earthquake of 1755: Representations and reactions* (Oxford: SVEC: 66-80).

9. Voltaire (1759) *Candid, or All for the Best* (London, J. Nourse).

10. Frèches, C. (1982) 'Pombal e la compagnie de Jésus: la compagne de pamplets', *Revista de História das Ideias,* 4/1: 299-327.

11. Diogo, M. (2001) 'Ciência Portuguesa no Iluminismo: Os estrangeirados e as comnunidades científicas europeias' in J. Nunes e M. Gonçlaves (eds.)*Enteados de Galileu? A semiperiferia no sistema mundial da ciência* (Porto: Afrontamento: 209-238).

12. Kagan, R. (2000) *Urban Images of the Hispanic World 1493-1793* (New Haven, Yale University Press).

13. Teixeira, M. C. and Valla, M. (1999) *O Urbanismo Português: Séculos XIII-XVIII : Portugal-Brasil* (Lisboa, Livros Horizonte).

14. Sepúlveda, C. (1910) *Manuel da Maya e os Engenheiros Militares Portugueses no Terromoto de 1755* (Lisbon: Imprensa Nacional).

15. Ratton, J. (1813) *Recordações de...sobre occurencias do seu tempo em Portugal durante o lapso de sestenta e tres annos e meio...*(Lisbon, Fenda 1992)

16. Manuel da Maia, *Dissertacão Part I, Part II and Part III* (1755-56) in J. França (1977) *Lisboa pombalina e o Iluminismo* (Lisboa: Bertrand).

17. Rossa, W. (2008) 'On the first plan' in A. Tostões. and W. Rossa *Lisboa 1758: The Baixa Plan Today* (Lisbon, Lisbon Municipal Council).

18. Rossa, W. (1998) *Beyond Baixa: Signs of Urban Planning in Eighteenth Century Lisbon* (Lisbon, Instituto Português do Património Arquitectónico).

19. Kostof, S. (1991) *The City Shaped: Uban Patterns and Meanings Through History* (London,Thames and Hudson).

20. Montiero, C. (2008) 'Laws written evenly along straight lines' in A. Tostões. and W. Rossa *Lisboa 1758: The Baixa Plan Today* (Lisbon, Lisbon Municipal Council).

21. Murteira, H. (1999) *Lisboa da Restauração às Luzes* (Lisboa: Editorial Presença).

22. Montiero, C. (2008) 'Laws written evenly along straight lines' in A. Tostões. and W. Rossa *Lisboa 1758: The Baixa Plan Today* (Lisbon, Lisbon Municipal Council).

23. Silva, R. (2008) 'Lisbon rebuilt and expanded 1758-1903' in A. Tostões. and W. Rossa *Lisboa 1758: The Baixa Plan Today* (Lisbon, Lisbon Municipal Council).

24. Subtil, J. (2007) *O Terramoto Político (1755-1759); Memória e poder* (Lisboa: Universidade Autónoma).

25. Almeida P. (1973) 'A arquitectura do século XVIII em Portugal: pretext e argumento para uma aproximacão semiológica', *Bracara Augusta* (Braga: Câmara Municipal de Braga 64/76 / XXVII / II:456).

26. Barreiros, M.H. (2008) 'Urban Landscapes: Houses, Streets and Squares of 18th Century Lisbon', *Journal of Early Modern History 12*, pp. 205-232.

27. Teixeira, M. C. and Valla, M. (1999) *O Urbanismo Português: Séculos XIII-XVIII : Portugal-Brasil* (Lisboa, Livros Horizonte).

28. McKinnon, M. (ed)(1997) *New Zealand Historical Atlas* (Auckland: David Bateman).

29. Canterbury Earthquake Recovery Act 2011: http://www.legislation.govt.nz/act/public/2011/0012/latest/DLM3653522.html

30. Christchurch City Council, Draft Central City Plan 2011: http://resources.ccc.govt.nz/files/CentralCityDecember2011/FinalDraftPlan/FinaldraftCentralCityPlan.pdf

31. Canterbury Earthquake Recovery Authority 2011: http://cera.govt.nz/government-response-to-the-august-2011-draft-of-the-central-city-plan

32. Christchurch Central Recovery Plan 2012: http://ccdu.govt.nz/the-plan

33. Davoudi, S., et al (2009) *Planning for Climate Change: Strategies for Mitigation and Adaptation*, London: Earthscan.

34. Earthquake Commission 2012: http://canterbury.eqc.govt.nz/faq

35. Muir-Wood, R. (2012) the Christchurch Earthquakes of 2010 and 2011 in Courbage, C. and Stahel, W. *The Geneva Reports-Risk and Insurance Research No 5, Extreme Events and Insurance, 2011 Annus Horribilis,* (Geneva: The Geneva Association).

36. Macfie, R. (2012) Frustration and Rage, *New Zealand Listener* September 18-14: 25-31.

Hazard Mitigation Planning in the United States: Historical Perspectives, Cultural Influences, and Current Challenges

Andrea M. Jackman[1] and Mario G. Beruvides[2]

[1]Advanced Analytics & Optimization, IBM Corporation, Pittsburgh, PA, USA

[2]Texas Tech University Department of Industrial Engineering, USA

INTRODUCTION

Planning for disasters at the federal, state, and local level is a relatively recent area of focus within the practice of emergency management in the United States. Historically, emergency management as a practice was focused on response to a disaster, with little attention paid to preparation, recovery, or overall and ongoing activities to reduce the effects of disasters. The theoretical framework and literature demonstrates the importance of planning as an activity which impacts the success of many other emergency management activities, yet practice has shown that planning is not always a valued or highly prioritized practice at the local level. The Disaster Mitigation Act of 2000 marked the first legislative emphasis on planning and mitigation and recent studies by the authors have shown mixed results for the implementation of planning laws. This chapter reviews in detail the historical developments in the theory and practice of planning with special emphasis on hazard mitigation planning; provides a theoretical framework based on the literature for understanding the importance of local level planning within the national system of emergency management, and the complexity that arises within that system; and discusses ongoing challenges in the successful completion of planning activities in the 21st century due to ongoing administrative and cultural challenges.

HAZARD MITIGATION BEFORE AND DURING THE COLD WAR

Understanding hazard mitigation in the United States first requires an understanding of how emergency management activities evolved historically. E. L. Quarantelli, one of the leaders in disaster sociology, described the beginnings of disaster research as "almost exclusively supported by the U.S.A. military organizations with very practical concerns about wartime situations" [1]. He notes that these "organized research activities [occurred] from about 1950 to 1965" and their primary goals were civil organization in wartime situations, under the assumption that "morale is the key to disaster control," and "effective disaster control includes the securing of conformity to emergency regulations" and "the reduction and control of panic reactions" [1]. The federal government took further action during the 1950s by undergoing several reorganizations within the Department of Defense (see [2]). Prior to, and during that time, the federal government was mainly concerned with civil defense, so that "private, voluntary agencies such as the American National Red Cross, the Salvation Army, and many others bore the primary responsibility for disaster relief; and state and local governments coped as best they could" [2]. Federal assistance was available as an absolute last resort by way of "special assistance acts passed by Congress" [2]. However this system had been operating essentially without change since 1803, and due to its reactive nature, there were "frequent delays before federal assistance reached impacted areas, and the nature of the assistance was designated only for selected purposes" [2].

Two interesting notes about the observations in [1] and [2]: first, the basis of government activity in emergency management emerged from a military and national defense perspective. The first "emergencies" in this regard were wars, or attacks from outside invaders. This militaristic approach – managing a disaster as enemy attack – would shape emergency management significantly in later years. Second, governmental activities in early years were largely reactive. Planning, particularly with an emphasis on mitigation, is not mentioned. A reactive war approach may seem antiquated outside of the Cold War context, but it is essential to understanding the development and decisions of current sentiments toward planning within local governments. As will be discussed in later sections, the defense mentality is still the dominant approach to loss prevention at the local level, and helps explain actions at all levels of government, in all modern aspects of emergency management.

THE FOUR PHASES OF EMERGENCY MANAGEMENT

In 1979, a report by the National Governor's Association was published on the topic of emergency management, and defined the general practice as:

The four phases listed- mitigation, preparedness, response, and recovery- serve as the current model of emergency management, are widely used among practitioners, and are considered the starting point for all policy and program design for all types of hazards at all levels of government. The NGA Report included only suggested actions for each phase, which were not operationally defined until 1985:

- Mitigation- assessing the risk posed by a hazard or potential disaster and attempting to reduce the risk;

- Preparedness- developing a response plan based upon the risk assessment, training response personnel, arranging for necessary resources, making arrangements with other jurisdictions for sharing of resources, clarifying jurisdictional responsibilities, and so on;

- Response- implementing the plan, reducing the potential for secondary damage, and preparing for the recovery phase; and

- Recovery- reestablishing life support systems, such as repairing electrical power networks, and providing temporary housing, food, and clothing. Recovery is assumed to stop short of reconstruction. [3]

In the years following the development of the NGA model, a number of scientific studies (summarized in Table 1) sought to define each phase in more detail. These definitions are still widely used today.

Table 1: Four Phase Model Definitions

Author	Preparedness	Response	Recovery	Mitigation
NGA Report, 1979 [4]	Developing a response plan and training first responders to save lives and reduce disaster damage, including the identification of critical resources and the development of necessary agreements between responding agencies	Providing emergency aid and assistance, reducing the probability of secondary damage, and minimizing problems for recovery operations.	Providing immediate support during the early recovery period necessary to return vital life support systems to minimum operation levels, and continuing to provide support until the community returns to normal.	Deciding what to do where a risk to the health, safety, and welfare of society has been determined to exist; and implementing a risk reductive program

Petak, 1985 [3]	[D]eveloping a response plan based upon the risk assessment, training response personnel, arranging for necessary resources, making arrangements with other jurisdictions for sharing of resources, clarifying jurisdictional responsibilities, and so on.	Implementing the plan, reducing the potential for secondary damage, and preparing for the recovery phase	Reestablishing life support systems, such as repairing electrical power networks, and providing temporary housing, food, and clothing	Assessing the risk posed by a hazard or potential disaster and attempting to reduce the risk
Comfort, 1985 [5]	Cities should review, exercise, and update their plans regularly based on staffing and past performance. Counties and states may review summarized local plans to identify resource needs and coordinate multijurisdictional exercises. FEMA may review state plans and adjust resources accordingly, as well as facilitate coordination between states.	Hierarchy proceeds from city, to county, to state, to federal. At the local level, responders make regular reports on status of life and property, assistance requests, at regular intervals. County, state, and federal designate aid, collect and analyze reports, summarize for next highest level and continue until basic systems are restored.	Assess damage and formulate short-term and long-term goals for rebuilding, including costs, needed equipment, and aid opportunities; ask for public input and improve rebuilt structures where possible; create schedule. All levels except city should identify and implement opportunities for inter-jurisdictional aid.	Conduct annual risk & vulnerability assessment with public involvement. Identify and formulate mitigation goals, and assign to appropriate agencies. County, state, and federal offices should monitor incoming reports and progress, allocate necessary resources, identify opportunities for inter-jurisdictional cooperation, and report to the next highest level.

| Waugh, 1990 [6] | Activities that develop operational capabilities for responding to an emergency (e.g. emergency operations plans, warning systems, emergency operations centers, emergency communications networks, emergency public information, mutual agreements, resource management plans, and training and exercises for emergency personnel | Activities taken immediately before, during, or directly after an emergency that save lives, minimize property damage, or improve recovery; e.g., emergency management plan activation, activation of emergency systems, emergency instructions to the public, emergency medical assistance, manning EOCs, reception and care, shelter and evacuation, search and rescue | Short-term activities that restore vital life support systems to minimum operating standards and long-term activities that return life to normal; e.g., debris clearance, contamination control, disaster unemployment assistance, temporary housing, and facility restoration. | Activities that reduce the degree of long-term risk to human life and property from natural and man-made hazards, e.g., building codes, disaster insurance, land-use management, risk mapping, safety codes, and tax incentives and discentives. |

FEMA, 2012 [7]	Actions that involve a combination of planning, resources, training, exercising, and organizing to build, sustain, and improve operational capabilities. Preparedness is the process of identifying the personnel, training, and equipment needed for a wide range of potential incidents, and developing jurisdiction-specific plans for delivering capabilities when needed for an incident.	Immediate actions to save lives, protect property and the environment, and meet basic human needs. Response also includes the execution of emergency plans and actions to support short-term recovery.	The development, coordination, and execution of service- and site-restoration plans; the reconstitution of government operations and services; individual, private-sector, nongovernmental, and public-assistance [housing and restoration] programs; long-term care and treatment of affected persons; [social, political, environmental, and economic restoration]; [identification of] lessons learned; postincident reporting; and development of [mitigation] initiatives	Activities providing a critical foundation in the effort to reduce the loss of life and property from natural and/or manmade disasters by avoiding or lessening the impact of a disaster and providing value to the public by creating safer communities... [F]ix the cycle of disaster damage, reconstruction, and repeated damage. These activities or actions... will have a long-term sustained effect.
Summary	*Threat assessment (TA) *Resource assessment & acquisition (RA&A) *Inter and intra-jurisdictional cooperation *Drills & Exercises (D&E) *Writing a plan (Plan)	*Activation of Emergency Protocol (AEP) *Medical assistance and first aid (EMS) *Shelter & Evacuation (S&E) *Search & Rescue (S&R) *Secondary Damage Reduction (SDR)	*Damage Assessment (DA) *Clean-up (De-con) *Restoration of critical systems & facilities (Restor) *Providing temporary basic needs (TBN) *Basic reconstruction (Recon I)	*Improved reconstruction (Recon II) *Legislative planning (LP) *Regularly scheduled vulnerability & risk assessments (VRA)

The four phases are widely considered to be overlapping and cyclical (Figure 1). Mitigation activities occur in all phases of a disaster, and frequently are most evident during reconstruction, which has since been informally added by practitioners as a part of the long-term recovery phase. The ongoing, ubiquitous nature of mitigation activities makes this the hardest phase to clearly define with a beginning and end point. As reconstruction and recovery near completion, lessons learned from these phases are incorporated into preparedness activities with additional mitigation in mind, which in turn are set aside when a response becomes necessary. Hazard Mitigation Plans are easiest to study within the context of the Planning phase, instead of Mitigation. According to the federal policy described later, mitigation, recovery and even some response activities are directed by state and local Hazard Mitigation Plans. Although risk assessment, defined here to be part of the mitigation, is a critical step in authoring a HMP, the entire process will be grouped into the Preparedness phase for simplicity. This is also due to the complex nature of risk assessment as a separate activity, and a tolerance for imprecision in the HMP approval process. Within the context of the Four Phase model, Preparedness, and specifically plan creation, at each level of government is described in the next section.

Planning for Disaster in Federal, State, and Local Government

The role of local-level emergency planning within the national emergency management framework is one of great importance. Federal government provides direction and goals for local planners, but primarily serves as a financial supporter when governments are unable to meet these goals. Likewise, the state acts as a regional conduit between federal and local government, providing aid to its local jurisdictions as needed. This concept, known as shared governance, is a reflection of American attitudes about self-governance. In their book exploring policy implementation issues within the federal government, May and Williams [8] cited, as an example of this mindset, the Elementary and Secondary Education Act of 1965, which marked the first time in U.S. history that the federal government assumed a direct funding role in public education. Although American government was deliberately designed in this fashion, it can cause a dilemma:

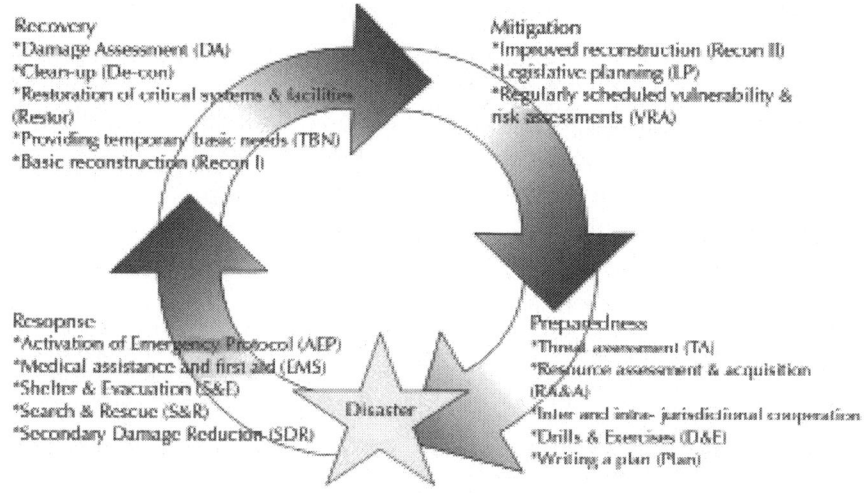

Figure 1: The cyclic nature of the Four Phase Model.

In the following sections, emergency planning at each phase of government will be discussed, with particular emphasis on local response to the recent federal demands for Hazard Mitigation Plans.

What is a hazard mitigation plan?

Before discussing how Hazard Mitigation Plans are completed within the government, it is worth briefly considering: what exactly is a Hazard Mitigation Plan? The Disaster Mitigation Act of 2000 [9] only lists two requirements for local mitigation plans, stating that the plans "shall (1) describe actions to mitigate hazards, risks, and vulnerabilities identified under the plan; and (2) establish a strategy to implement those actions" [P.L. 106-390 § 322(b)]. FEMA's Interim Final Rule (The Rule) provides much more specific requirements based on these guidelines. In summary, a Hazard Mitigation Plan must include:

- Documentation of the planning process;
- A risk assessment, including: (i) a description of the type, location, and extent of all natural hazards that can affect the jurisdiction, including previous occurrences and (ii) a description of the jurisdictions vulnerability to the hazards. Vulnerability should be described in terms of: (A) types and numbers of

existing infrastructure, (B) an estimate of potential dollar losses to vulnerable structures, and (C) a description of land uses and development trends. (iii) "For multi-jurisdictional plans, the risk assessment section must assess each jurisdiction's risks where they vary from the risks facing the entire planning area."

- A mitigation strategy, including: (i) long-term mitigation goals, (ii) a description of specific actions for new and existing structures, and (iii) an action plan for how the above will be implemented, prioritized by cost-benefit analysis.

- A plan maintenance process, including: (i) a description of maintenance for the plan on a five-year cycle, (ii) a process, if possible, to incorporate mitigation efforts into other aspects of local planning, and (iii) a discussion on continuing public maintenance of the plan.

- Documentation that the plan has been formally adopted by all participating jurisdictions [44 CFR 201.6(c)].

Because the legal style of The Rule can be tedious and lacking examples, FEMA published a series of how-to guides for state and local mitigation planning [10]. The first four guides listed are considered the "Core Four" of HMPs, with the remaining guides available for those jurisdictions as applicable:

- Getting started with the mitigation planning process, including important considerations for how you can organize your efforts to develop an effective mitigation plan (FEMA 386-1);

- Identifying hazards and assessing losses to your community, State, or Tribe (FEMA 386-2);

- Setting mitigation priorities and goals for your community, State, or Tribe and writing the plan (FEMA 386-3);

- Implementing the mitigation plan, including project funding and maintaining a dynamic plan that changes to meet new developments (FEMA 386-4);

- Evaluating potential mitigation actions through the use of benefit-cost review (FEMA 386-5);

- Incorporating special considerations into hazard mitigation planning for historic properties and cultural resources, the topic of this how-to guide (FEMA 386-6);

- Incorporating mitigation considerations for manmade hazards into hazard mitigation planning (FEMA 386-7);
- Multi-Jurisdictional Mitigation Planning (FEMA 386-8); and
- Finding and securing technical and financial resources for mitigation planning (FEMA 386-9).

All of the guides have a similar format of listing the specific subsection of The Rule, and then provide an explanation, a list of required activities, recommended activities, and examples for how to implement the specific part of The Rule in a clear, non-legal style.

The eighth volume of the How-To Guide, published in 2006 (386-8), is titled "Multi-Jurisdictional Mitigation Planning" and provides guidelines for this specific type of local plan authorship. Although there are many ways to organize a multi-jurisdictional plan, the guide recommends a specific structure to follow; the common portion of the plan may include the "process, common hazards, general mitigation goals, collaborative actions, and [plan] maintenance [schedule]." The items unique to each participating jurisdiction that may be included are: "geographically specific hazards, risks, specific [mitigation] goals, actions, participation, and adoption" [10]. In other words, the number of activities for which the costs would fall exclusively to a single jurisdiction has already been reduced.

If a plan is to be submitted as a multi-jurisdictional HMP, 386-8 provides specific requirements that must be met at each stage of the process. FEMA 386-8 makes recommendations for how to implement the requirements, and tips and examples for following the recommendations. Since the recommendations are not mandatory, and each jurisdiction is unique, the recommendations are not included in summary table. One critical component for multi-jurisdictional plans however, is "documentation" or "proof or adoption" is required from participating single jurisdictions. This refers to city or county resolutions that were passed in the individual jurisdictions to adopt the regional or multi-jurisdictional mitigation plan.

With regard to plan participation, the organization of multiple jurisdictions generally follows three models: Direct Representation, Authorized Representation, and a combination of the two. The first involves sending "direct representatives" to the plan author, who coordinates the creation of the plan. For the second, the individual jurisdictions will authorize the plan author to act on their behalf, usually through city or county resolution [10]. A combination of the two can also be created. Any or all of the models are acceptable, but may lead to different cost situations.

Planning At the Federal and State Level

As the U.S. exited the Cold War, emergency management at all levels of government continued to evolve and in 1974 with The Disaster Relief Act was enacted. The primary goal of the Disaster Relief Act was to update the federal response and relief system described earlier, and to grant more power to the federal government to provide aid in the immediate aftermath of a disaster. In 1979, following the Disaster Relief Act, the Federal Emergency Management Agency (FEMA) was formed. While FEMA remains the national organization for emergency management, past structuring of the federal bureaucracy has shown that these institutions are frequently replaced. Predecessors to FEMA include: The Office of Civil and Defense Mobilization (1958), the Office of Emergency Preparedness (1961), The Civil Defense Preparedness Agency (1972), and finally the Federal Emergency Management Agency in 1979 (see [2]). Each of these contained multiple sub-organizations concerned with different areas of emergency management, and operated within a wide range of government groups, from the Department of Defense (DOD) to Housing and Urban Development (HUD) [2]. As a result of the terrorist attacks on September 11, 2001, FEMA was brought under the auspices of the newly created Department of Homeland Security (DHS); and after a controversial response to Hurricane Katrina in 2005 CNN reported that a congressional committee was calling for the abolition of FEMA [11].

After the changes made at the federal level during 1970s, policy continued to evolve through amendments to the Disaster Relief Act of 1974 with the Robert T. Stafford Disaster Relief and Emergency Assistance Act (1988), and the Disaster Mitigation Act (2000). Each amendment encourages localities to "focus on individual and community infrastructures," unless the disaster is beyond their ability to manage [12]. Further, "if the disaster exceeds the state's capacity to respond ... the state governor [is allowed] to request aid from the national government. FEMA evaluates the request, prepares material for presidential approval, and coordinates the federal response" [12]. Local and state governments now officially bore the responsibility for emergency planning, although federal response capacity had been expanded.

The Disaster Mitigation Act of 2000 was significant because by its own title was the first law to emphasize the mitigation and preparedness phases of the Four Phase model, rather than "relief" or "assistance" as before; this was achieved by expanding Section 404 of the Stafford Act, which authorized the Hazard Mitigation Grant Program (HMGP) as a means by which jurisdictions *that had received presidential declarations of disaster* could apply for and receive federal assistance for mitigation projects. An additional program, for Pre-Disaster Mitigation grants (PDMs), was instituted so that a presidential declaration was not a requirement to apply for funding directed at mitigation activity; however the application process is separate, nationally competitive, and less familiar than that of the HMGP; and often the amount of money made available for funding applications through presidential declarations is substantially higher. In amending Section 404 of the Stafford Act, Section 322(a) of the Disaster Mitigation Act required state and local mitigation plans to be in place before any applications were made to the HMGP:

The Disaster Mitigation Act provided a legal foundation for FEMA to author an Interim Final Rule under the Federal Register (44 CFR Parts 201 and 206). As discussed in the previous section, the Rule provides specific clarification, based on the Disaster Mitigation Act, for receiving funding through FEMA under the HMGP. Beginning at the state level, a state can either have a Standard or Enhanced Mitigation Plan that will result in a 15% or 20% increase in HMGP funding, respectively. The state is also allowed to use up to 7% of the HMGP funding to cover the expenses of writing state, local, or tribal plans. As of November 2007, 48 states had approved Standard Plans, and two states were waiting for approval on submitted plans. Seven of the 48 states with approved plans had also elevated their status to having approved Enhanced Plans, showing the state-level implementation of plans was highly successful. The Rule explicitly states that "[t]o be eligible to receive HMGP project grants, local governments must develop Local Mitigation Plans that include a risk assessment and mitigation strategy to reduce potential losses and target resources. Plans must be reviewed, revised, and submitted to us for approval every 5 years" (p. 8847). Local Mitigation Plans are also referred to as Hazard Mitigation Plans (HMPs), or Mitigation Action Plans, by FEMA and local planners alike. An important note for later discussions on the cultural influences in local planning, The Rule further specifies that "[m]ulti-jurisdictional plans may be accepted, as appropriate, as long as each jurisdiction has participated in the process and has officially adopted the plan. State-wide plans will not be accepted as multi-jurisdictional plans" [44 CFR § 201.6(3)].

To encourage a fast response to the new local-level planning requirements, The Rule originally set a deadline of November 1, 2003. Prior to that date, writing plans and applying for funding through the HMGP could be done simultaneously. In October 2003 the deadline was changed to November 1, 2004 with an amendment in the Federal Register, stating that "local governments must have an approved mitigation plan in order to receive project grants under any Notice of Funding Opportunity [including PDMs] issued after November 1, 2003 [fiscal year 2004 and later]" (p. 61368). Interestingly, this legislation used a limitation of access to federal grants to motivate local governments to create HMPs.

From this sequence of bureaucratic re-organization and policy implementation, it is clear that planning for disasters at the federal level has involved maintaining a reliable response and relief capacity, and passing the planning responsibilities to state and local government. This is not counterintuitive however, as local residents have a better understanding of their areas, and would be the first to respond during a disaster.

Planning at the Local Level

While federal and state governments are easily recognizable, it is worth considering the definitions of local government when considering the planning that occurs there. The U.S. Census Bureau provides rigorous definitions for city governments, and a certain set of criteria that must be met for a local government to be considered legitimate. FEMA accepts plans from a wide variety of local governments, including tribal governments and individual school districts. When conducting any analysis on HMPs, a distinction should be made for which types of governments are under consideration. Councils of governments are not defined by the census bureau, and may take a variety of forms depending on the needs of localities within a region. According to the

National Association of Regional Councils (NARC), a regional council, or council of governments, is defined as:

In support of the notion within emergency management that inter-organizational cooperation is crucial, [13] believes "the role of the regional council has been shaped by the changing dynamics in federal, state and local government relations, and the growing recognition that the region is the arena in which local governments must work together to resolve social and environmental challenges."

As emergency management evolves and becomes more advanced, the earlier quotation from [8] becomes more relevant. Recall that:

Because of increased globalization, a community that was once relatively isolated might now house critical facilities for a distant parent company. Sociologist Arjen Boin notes how deeply systemic and interlinked society as become, allowing the effects of disaster to spread and multiply more rapidly than in the past, and stressing the need for improved local disaster planning:

All levels of government participate in some way in all levels of emergency management, creating a complex system of interlinked activities. Ultimately though, the entire structure of emergency management in the United States, and within the Four Phase model, depends on preparedness at the local level. This concept is aptly publicized by the planning requirements within the Disaster Mitigation Act and FEMA's Interim Final Rule. Despite general consensus that local preparedness is essential, its execution has traditionally been of minimal quality, low priority, and host to a multitude of administrative problems. These are discussed in the following sections.

What constitutes preparedness?

Returning to the Four Phase model of emergency management proposed in 1979 by the NGA, the report failed to provide definitions for the phases; instead, suggested activities were included. For the preparedness phase, the NGA recommended:

Six years later, the NGA was better able to define each phase (see Table 1). Preparedness was defined as:

An interesting similarity between both definitions is that they encourage cooperation with other jurisdictions. Although this cooperation has appeared low on the list of priorities of local planners for reasons discussed later, recent research has shown multi-jurisdictional cooperation to be almost exclusively responsible for the creation of HMPs [15].

As the understanding of emergency planning and hazards progressed, a number of researchers would recommend activities that led to an increased state of preparedness for local emergency managers (see [16]). After the terrorist attacks of September 11, 2001, [16] revisited these activities, summarized and combined the work that had been done previously, and suggested ten guidelines for increased preparedness within the newfound context of terrorism as a viable threat. In summary, the ten steps are:

- Base planning activities "upon accurate knowledge of the threat and of likely human responses;"
- encourage an appropriate, rather than quick or impulsive, response;
- emphasize "response flexibility so that those involved in operations can adjust to changing disaster demands;"
- address inter-organizational coordination;
- "integrate plans for each individual community hazard managed into a comprehensive approach for multi-hazard management;"
- include a training program so that all involved parties are familiar with the plan, including elected officials and the general public;
- test the plan with drills and exercises;
- recognize that "planning is a continuing process;"
- recognize that due to the nature of local government culture [see Section 2.2.3.2.3], "emergency planning... is almost always conducted in the face of conflict and resistance;" and
- 1recognize that a plan is only ever truly tested and improved upon "with its implementation in an emergency" (adapted from [16]).

The authors note that "often, there is a tendency to equate emergency planning with the presence of a written plan and similarly believe that a written plan is evidence of jurisdictional preparedness" [16]. In fact, as demonstrated in the ten guidelines, planning is a dynamic process. Emphasizing a written plan may not be a bad idea, given the requirements of the Hazard Mitigation Grant Program; however a possible future task for policy might be to highlight the process rather than the written document.

Combining the definitions of the NGA Four Phase model with [4] and [16], preparedness within the context of emergency management is best thought of as a cyclic process, much like the Four Phase model, which consists of threat assessment, resource assessment and acquisition, inter- and intra-jurisdictional cooperation, drills and exercises, and finally writing a plan (see Figure 1). As previously discussed, a preliminary examination of FEMA data on Hazard Mitigation Plan completion has shown that over 90% of the "plan writing" phase of preparedness has occurred at the multi-jurisdictional level, especially within counties and COGs [15]. It would appear that these five activities within preparedness can occur with varying success at different levels of local government. The history of multi-government bodies in emergency management is discussed in the next section.

The Role of Counties and Councils of Governments

With rare exception, emergency management literature has followed the governmental design of the NGA model to the letter; the four phases are to be carried out at the federal, state, and local level. However, in the NGA report and subsequent literature, local government is seldom defined and assumed to mean primarily city, or occasionally, county government. Very little literature exists on the role of councils of governments in the preparedness phase.

An important note from the literature in emergency management is that "inter-organizational" or "multi-jurisdictional" coordination is considered essential among disaster researchers; even if the terms are broad, encompass many types of coordination, and refer almost exclusively to the response phase of emergency management. Like [14], Louise Comfort argues that due to the increasing complexity of society, not only are effective local responses critical, but are also "necessarily inter-organizational and interdisciplinary" [17]. Comfort had previously proposed specific roles for county emergency management within the preparedness phase. In summary, Comfort lists the county's responsibilities as:

- Review individual city emergency plans and enter their data into a resource database;
- Summarize database into county-wide profile of responsibilities and capabilities, and return this report to city governments for review;
- Conduct drills and exercises that bring multiple organizations together; 4) evaluate the performance of the cities in these drills;
- improve preparedness at the county level and "seek assistance... from inter-jurisdictional sources;"
- schedule, monitor and evaluate preparedness activities; and
- submit an annual report of these activities to the state (adapted from [17]).

Two important factors in Comfort's guidelines are that first, she recognizes the importance of a coordinating government to act between the city and state levels, but she also relies on the assumption that individual cities will author their own plans.

In 1994, William Waugh expanded on Comfort's role for county government. Waugh argued that counties should be the exclusive home of local emergency management, because county offices generally:

- are geographically close to environmental problems,
- have larger resource bases than municipalities,
- have ambiguous administrative structures that encourage inter- and intra-organizational cooperation,
- are local agents of state administration,
- have close administrative ties to state agencies,
- provide forums for local-local cooperation, and
- serve as general-purpose governments representing local interests and have strong local identification (adapted from [3]).

Waugh's reasoning may provide some insight into why the success rates for Hazard Mitigation Plan authorship are so high for counties and COGs. Yet in many rural areas, counties only encompass a small number of sparsely populated municipalities, which raises the question of when county governments or COGs are more appropriate in the planning process.

Only one example of a successful COG exists in the literature, and it receives a brief mention in a report by Thomas Drabek [18]. In 1990,

Drabek published the results of a study of twelve highly successful local emergency managers. From what he learned through personal visits and interviews, Drabek extracted fifteen qualities that all of the managers shared; one of which was the formation of "mergers." While this generally meant the cooperation between public and private organizations, or inter-departmental cooperation, Drabek found that Donald Herrick of Davidson County, South Dakota founded the James Valley Emergency and Disaster Service District- "a four county emergency services unit" [18].

Undoubtedly the academic aspect of emergency management recognizes the usefulness of regionalized government, especially counties and within the response phase of a disaster. In practice at the local level however, both the preparedness phase of emergency management and the concept of shared governance even at a regional level is resisted and viewed with suspicion and disdain. Despite its apparent benefits, the difficulty in implementing multi-jurisdictional cooperation is discussed next.

Cultural Issues in Local Government

Planning for disaster in local government has traditionally been a neglected and misunderstood part of emergency management. The reasons, summarized and listed in [19], include:

In other words, emergency management is not a simple matter. The complex and infrequent nature of disasters compared with more familiar problems places them low on the list of priorities for many planners. This lack of enthusiasm is compounded by local politics, turf protection, and ambiguity caused by shared governance. These reasons for resistance to planning efforts can cause both vertical and horizontal fragmentation of government.

Documenting this type of cultural phenomenon poses a challenge of a sociological nature. Presented below are the results of preliminary studies that have begun quantifying these barriers to success. The results indicate that an aversion to planning is frequently present among local government officials. The reason is twofold: the process itself is ongoing, expensive, and time-consuming, and the background of many professionals in emergency management is one of trained rapid response. By asking city planners to rate their own successes in the formation of mandated local toxic chemical emergency planning committees (LEPCs) under SARA Title III, five years after the policy went into effect in the state of Michigan, M. Lindell [20] found that:

Lindell's results indicate that not only are planners reluctant to take action, but willingly rank themselves as such. In a follow-up study [21], Lindell found that the largest contributors to the time commitments needed for plan completion were: committee member input, available planning resources, and community support. Staffing and structure within the government and the city's vulnerability to hazards were not found to be significant (see [21]).

Lindell's findings [20, 21] were supported by two recent papers (see Buckle et al. [22]; and Stuart-Black et al. [23]). Buckle et al. found that the unfamiliar nature of hazards made them less appealing for planners, and that good communication between local government and community led to better planning [22]. The second study [23] surveyed local emergency managers to determine the composition of the field with regard to education, background, age, sex, and previous job experience. The results demonstrated a lack of value placed on education or academic training, with preferences given to practical experience in defense or response-oriented jobs. One of the motivations for the study was what the authors described as an informal "notion... that those doing the job were older men from a military or emergency services background, who having retired from their service were embarking on a second career in order to boost their pensions" [23]. In the United Kingdom, the study found that 76% of local planners looking to hire a new emergency manager were not even considering recent graduates or degree holders [23]. The planners estimated they would fill their positions using employees with significant experience or those looking for a transition into retirement. When asked where they expected to find potential candidates, the planners responded that they "expected to recruit from the local government sector (63%), first response (37%), and/or retired military (34%)," with percentages including responses where multiple sectors were chosen as potential hiring pools. The surveys also asked why these sectors where chosen, and "the overwhelming answer was that age and experience were paramount to the job, and younger applicants were not always able to bring the necessary authority that was needed in dealing with senior officers and elected council members." In regard to this "overwhelming" response, the authors commented that "clearly the emergency planners are by their own actions and beliefs perpetuating the myth." Though the "notion" that prompted surveys in [23] was informal and not fully documented, it certainly is supported by the data collected.

Local emergency managers appear to subscribe to the war-oriented approach described by [1] above. Often police and fire departments closely resemble the military in structure, training, and operation, with all groups placing high emphasis on the ability to act rationally and maintain order in emergency situations. As indicated by [23], this leads directly to hiring preferences that value the experienced responder above all other candidates. It also leads to a second inhibitor to local planning: the difficulties of implementing inter-jurisdictional cooperation.

Policy research has shown that because of differing priorities of various agencies, such as police and fire, "bureaucrats tend to avoid communication with their counterparts in other agencies, even when their responsibilities clearly overlap or interface... In general, the more coordination required to implement a policy, the less chances of its success" (Edwards, 1978, as quoted in [24]). Kartez and Kelley [25] supported this finding with their own survey of local emergency planners. The planners were asked to rank seven strategies for implementing preparedness policy, based on perceived likelihood of adoption, perceived benefits of strategy, and perceived effort of adoption. Among other strategies, such as citizen education and creating a media information center, inter-jurisdictional forums ranked third and second respectively in benefit and effort, but dropped to fourth for the likelihood of adoption [25]. The authors surmised that the planners recognized the benefit of inter-jurisdictional collaboration, but deemed it too difficult to execute.

Drabek's study [18] of successful emergency managers also supported these conclusions, highlighting the political reasons for avoiding working with other jurisdictions and even departments within their single jurisdiction. Drabek sited "turf defense" as a major barricade to what he called the "sensitive ground" of "coalition building" [18]. Drabek specifically cited an emergency manager that had tried to start a smoke detector and fire extinguisher campaign in his jurisdiction, much to the irritation of the fire department, who felt such a campaign was their responsibility and resented the emergency manager for making them look unconcerned about prevention.

Summary of Planning For Disaster in Federal, State, and Local Government

The previous sections provided a history of the planning subsection of the preparedness phase of emergency management. Planning at the federal level is limited; federal government is primarily a financier and supporting partner of response, recovery, and mitigation efforts. The most recent federal policy, the Disaster Mitigation Act of 2000 and FEMA's subsequent Interim Final Rule (44 CFR Parts 201 and 206) have required that all local jurisdictions have an approved Hazard Mitigation Plan in order to be eligible for any federal funding opportunities.

The states play intermediate roles in transferring information between local and federal governments, and the local governments are responsible for their own planning. Using the five aspects of preparedness [4, 16], Table 2 shows how some roles within the Preparedness phase can be checked off by definition, while others remain poorly understood. To carry out any of the activities listed at the regional level, without the knowledge or cooperation of the city level, would be extremely poor planning. Similarly, inter- and intra-jurisdictional cooperation requires the participation of multiple jurisdictions by definition. The remaining roles however, are poorly understood within the literature. For instance, what is the role of a council of governments in drills and exercises? Are they activities that require maximum cooperation, or are counties better suited to perform this task so as to avoid over-complication? This chapter focused on planning at the city, county, and COG level (Table 2) but certainly more research is needed in the other areas of the Preparedness phase.

Table 2: The Roles of Local Government within the Preparedness Phase

COG	?	?	?	+	?
County	?	?	?	+	?
City	+	+	+	+	+
	Plan	TA	RAA	IJC	D&E

Although placing responsibility for planning at the local level is logical, considering locals know their areas the best and are the first to respond to a disaster, literature shows that in practice there are many more factors at play. First, writing a plan on paper is only a small portion of preparedness as a whole. Second, the individual government success rate for Hazard Mitigation Plans is minute compared to that for multi-jurisdictional bodies; even though the latter is not well understood in the literature. Finally, a political, response-based culture at the local level has consistently made multi-jurisdictional cooperation difficult.

Returning to planning and preparedness within the context of a national emergency management system, recall that emergency management follows a four-phase model developed in 1979 by the National Governor's Association. The four phases are: preparedness, response, recovery, and mitigation. They are accepted as standard among practitioners of emergency management, and are widely considered to be overlapping and cyclical (Figure 1). All four phases contain component activities as demonstrated in the literature (Table 1). Due to the complexity of actual disasters, it is likely that even more activities and sub-categories exist within these divisions, but they have yet to be formally established by the literature.

As defined by the NGA model, the four phases of emergency management can be extended to all levels of government (Figure 2). A typical assumption in emergency management literature is that government in the United States is divided into local, state, and federal levels. However local government can be further subdivided into municipality/town, county, and COG. The activities that comprise the four phases of emergency management may be carried out at all levels of government.

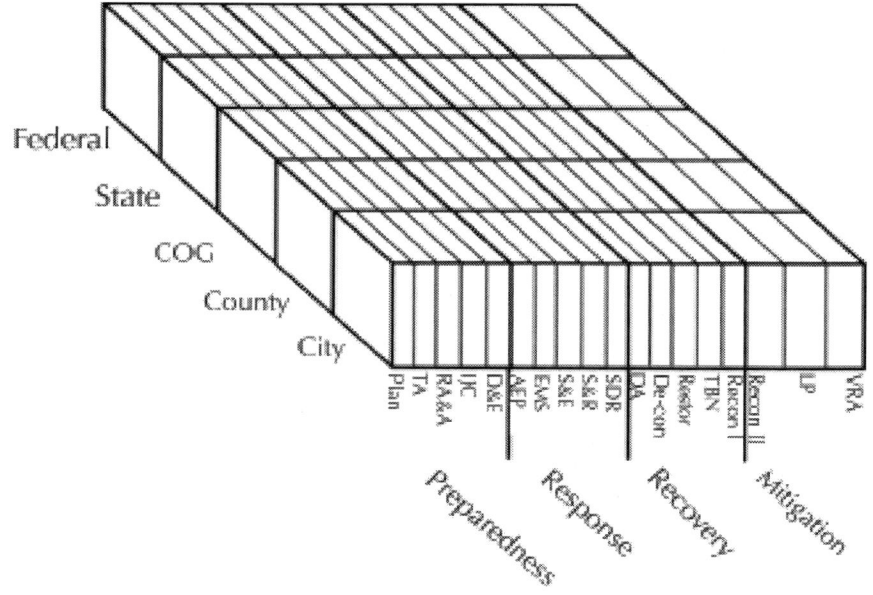

Figure 2: Four phases of emergency management at all levels of government.

However a third dimension may be added to the model to show what aspects of emergency management can influence the activities within certain areas of government. Three factors were found to have a significant effect on organizing emergency management activities within a government by [21] as discussed earlier: available resources, committee input, and community support. It is likely that there are many more factors that influence preparedness and cooperation in local emergency planning, but these have yet to be documented in the literature. In addition to influencing emergency management activities, these three factors also provide frameworks for measuring the activities. A pictorial representation (Figure 3) provides a visual summary of the Four Phase Model, extended to all levels of government, and within the contexts for action identified by [21], and clearly shows the complexity faced by local planners.

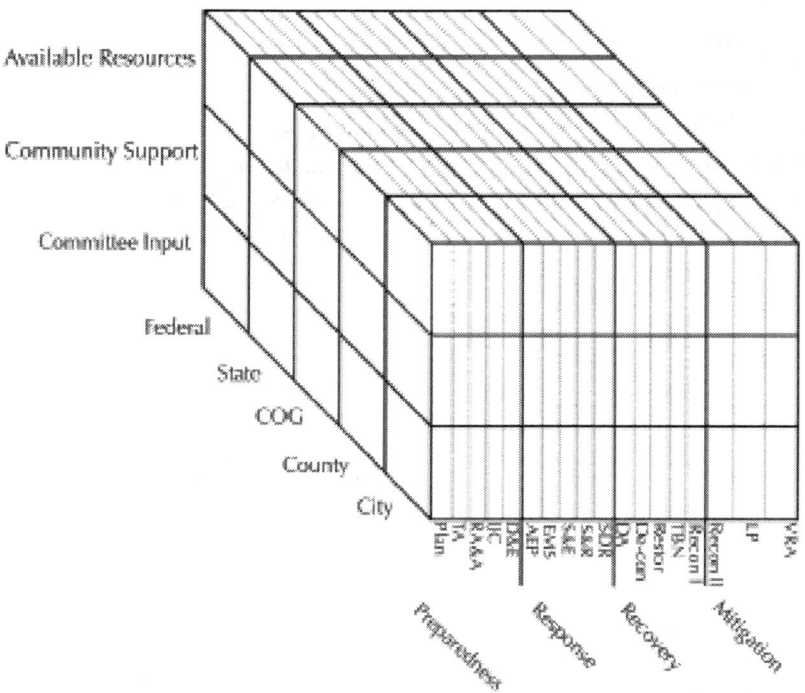

Figure 3: A Conceptual Model of the National Emergency Management System. Copyright © 2008 Andrea M. Jackman & Mario G. Beruvides.

Hazard Mitigation Planning as Part of a National Emergency Management System

Most of the emergency management literature in this chapter is presented within the context of planning, specifically for hazard mitigation in a local community. However based on Table 1, the findings of [21], and the established structure of American government, it is not unreasonable to begin imagining the complexity of our national emergency management system as illustrated in Figure 3. Certainly there is more research to be done; more activities may be added to the subdivisions of the Four Phase Model as our national approach to emergency management grows and evolves, and further motivating factors for each activity will likely be discovered beyond those in [21] that were found to influence planning.

However one aspect of Figure 3 cannot be disputed: the complexity of our national emergency management system will not get any simpler. Even the introductory overview of literature provided in this chapter is able to justify an 18 x 5 x 3 conceptual diagram – equaling a minimum of 270 individual components that make up the national system of emergency management. Recalling the words of sociologist Arjen Boin from earlier:

Hazard mitigation planning is a small component of emergency management. Even expanded to all possible levels of government, it is only one type of plan among many, and planning is only one type of activity in overall preparedness. Yet one might question, how "tightly coupled" is it with other aspects of emergency management? How rapidly will one action within a HMP propagate to other subsystems within emergency management as a whole? A simple HMP may be comprised of a community risk assessment and one or two mitigative actions to reduce those risks. But the risk assessment is likely based on past disasters in the community. The lessons learned and recommended actions from those disasters in turn influence future responses, which influence future recovery efforts, which will drive mitigation planning and risk assessments in later years. Through Figure 3 we see how one activity affects many others within the system. At first glance, local hazard mitigation planning seems distant and unrelated to decontamination efforts managed by the federal government. However an effective mitigation strategy put in to place today through the HMP process may significantly reduce the need for decontamination or any federal involvement at all. As another example, the after-action report of a state-level search and rescue team could directly impact risk assessments, planning, and mitigation strategy following a major disaster.

Hazard mitigation planning at the local and COG level, studied from all possible planning contexts, only comprises (at most) 9 out of 270 subsections of Figure 3, or 3%. This estimate does not include the further breakdown of different kinds of plans in addition to HMPs, yet was shown to influence many other subsections of Figure 3. This illustrates not only the importance of understanding hazard mitigation plans, but the impact of *any legislative action* taken in emergency management. The true impact of a single act can have vast, sometimes unpredictable consequences, especially in a system such as emergency management where current practices and scientific research are still relatively new. An understanding of the implementation of the HMGP policy is critical for this reason, and is discussed in the next section.

HAZARD MITIGATION PLANNING IN THE 21ST CENTURY

The HMGP policy that led to HMPs as a requirement was put into place in November, 2004. Based on the material covered in the previous sections, two questions naturally arise: first, how many local jurisdictions have completed HMPs since the original deadline? Second, for those localities with an approved HMP, how did they manage given all the documented cultural aversions to planning at the local level?

These questions were answered in part by a recent series of studies [15, 26]. An initial study [15] found that in 2008, 67% of the country's active local governments were without an approved Hazard Mitigation Plan (Figure 4).

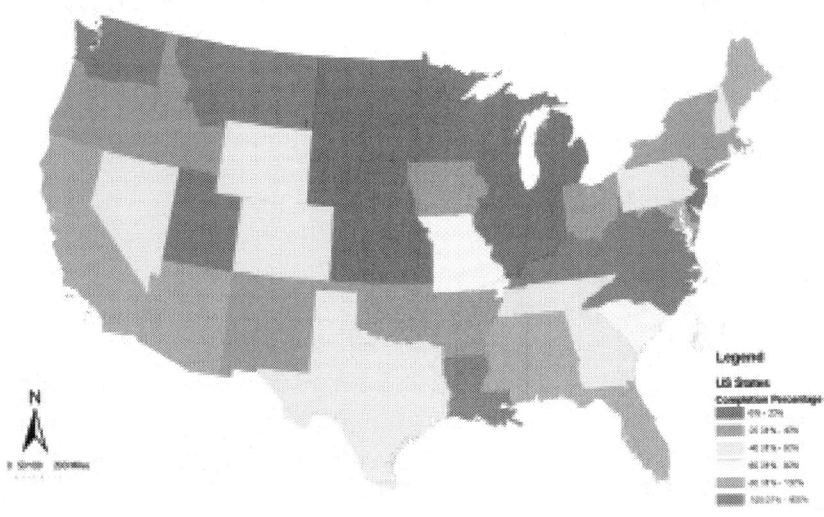

Figure 4: Map of Hazard Mitigation Plan Completion Percentage for the Continental United States in 2008 [15].

A follow up examination in 2009 [15] of the eight states with the lowest completion percentages did not indicate significant improvement following the initial study, and revealed inconsistencies in plan completion data over time. The completion percentage varied greatly by state, and did not appear to follow any expected pattern such as wealth or hazard vulnerability that might encourage prompt completion of a plan. Further, the results indicated that approximately 92% of the approved plans were completed by multi-jurisdictional entities, which suggests single governments seldom complete and gain approval for plans. This is directly opposed to expectations set by literature documenting cultural barriers to multi-jurisdictional collaboration, and presents a number of opportunities for further research.

The study was conducted for the initial three year period of the HMGP from 2004 to 2007, and given the results, it is important to note that federal policy such as the HMP requirements can change quickly and often. Strategic directions, policy, and guidance can change regularly, and is always expected at the federal level following a change in administration. The completion percentages demonstrated in this study represent an important step in understanding how long it takes for jurisdictions to react to policy changes and take necessary steps to become compliant, especially given the systemic complexity demonstrated in Figure 3.

A second study [26] examined HMP completion within the context of "available resources" fromFigure 3; namely, cost. It was found that the cost of a HMP varied significantly based on the frequency of natural hazards experienced by the authoring jurisdiction, the number of participating jurisdictions in the plan, population, and population density. Similarly, multi-jurisdictional plans were found to be significantly cheaper unless a jurisdiction experienced, on average, more than 6.5 events requiring some kind of emergency response per year (see [26]). This would provide a financial incentive for jurisdictions to override some of the cultural barriers mentioned earlier, and proceed with a multi-jurisdictional plan. In view of the realities presented thus far and the sheer complexity of the US emergency management system, future research might benefit from a systems analysis and systems dynamic modeling to assist in shaping our national emergency management policy.

Where will hazard mitigation planning go from here? The importance of having at least some level of understanding of the possible impacts of any new emergency management policy were illustrated by Figure 3, and this section demonstrates that for the example of hazard mitigation planning, relatively little is known about its implementation, success, and longevity. Planning in general was shown by the literature to be valued by policymakers and theorists, but difficult to execute in practice for a variety of reasons. Due to the far-reaching consequences of good mitigation and mitigation planning, continued research in this area is critical to a better understanding of our entire national emergency management system.

ACKNOWLEDGEMENTS

The research presented in this chapter was originally conducted as a part of the first author's doctoral thesis. See [27].This research was funded in part by the National Science Foundation, Grant No. 022168.

REFERENCES

1. Quarantelli EL. Disaster Studies: An Analysis of the Social Historical Factors Affecting the Development of Research in the Area. International Journal of Mass Emergencies and Disasters 1987; 5(3), 285-310.

2. Kreps GA. The Federal Emergency Management System in the United States: Past and Present. International Journal of Mass Emergencies and Disasters 1990; 8(3), 275-300.

3. Waugh WL. Regionalizing Emergency Management: Counties as State and Local Government. Public Administration Review 1994; 54(3), 253-258.

4. Petak WJ. Emergency Management: A Challenge for Public Administration. Public Administration Review 1985; Special Issue, 3-7.

5. Comfort LK. Integrating Organizational Action in Emergency Management; Strategies for Change. Public Administration Review 1985; Special Issue, 155-164.

6. Waugh WL, Hy RJ, eds. Handbook of Emergency Management: Programs and Policies Dealing with Major Hazards and Disasters. Westport, CT: Greenwood Press; 1990.

7. Federal Emergency Management Agency. National Response Framework Resource Center: Glossary/Acronyms. http://www.fema.gov/emergency/nrf/glossary.htm (accessed 19 February 2012).

8. May PJ, Williams W. Disaster Policy Implementation: Managing Programs under Shared Governance. New York, NY: Plenum Publishing Company; 1986.

9. Disaster Mitigation Act of 2000, Pub. L. No. 106-390, 114 Stat. 1552.

10. Federal Emergency Management Agency. How-To Guide for State and Local Mitigation Planning (No. 386). Jessup, MD: Author; 2002.

11. American Morning [Television News Program]. Should This Be the End of FEMA? (27 April 2006). Lexis-Nexis at transcripts.cnn.com (accessed 21 May 2007).

12. Daniels RS, Clark-Daniels CL. Vulnerability Reduction and Political Responsiveness: Explaining Executive Decisions in U.S. Disaster Policy during the Ford and Carter Administrations. International Journal of Mass Emergencies and Disasters 2002; 20(2), 225-253.

13. National Association of Regional Councils. What is a Regional Council? http://narc.org/regional-councils-mpos/what-is-a-regional-council.html (accessed 20 November 2007).

14. Boin A. Disaster Research and Future Crises: Broadening the Research Agenda. International Journal of Mass Emergencies and Disasters 2005; 23(3), 199-214.

15. Jackman AM, Beruvides MG (2008). Local Hazard Mitigation Plans: A Preliminary Estimation of State-Level Completion from 2004 to 2009. Working paper; accepted for publication on 9 August 2012 to the Journal of Emergency Management.

16. Lindell MK, Perry RW. Preparedness for Emergency Responses: Guidelines for the Emergency Planning Process. Disasters 2003; 27(4), 336-350.

17. Comfort LK. Designing Policy for Action: The Emergency Management System. In: Comfort, LK. (ed.) Managing Disaster: Strategies and Policy Perspectives. Duke Press: Durham, NC: Duke Press; 1988.

18. Drabek TE. Emergency Management: Strategies for Maintaining Organizational Integrity. Ann Arbor, MI: Springer-Verlag; 1990.

19. Waugh WL. Emergency Management and the Capacity of State and Local Government. In: Sylves RT, Waugh W (eds.) Cities and Disaster: North American Studies in Emergency Management. Springfield, IL: Charles C. Thomas; 1988.

20. Lindell MK. Are Local Emergency Planning Committees Effective in Developing Community Disaster Preparedness? International Journal of Mass Emergencies and Disasters 1994; 12(2), 159-182.

21. Lindell MK, Meier MJ. Planning Effectiveness of Community Planning for Toxic Chemical Emergencies. Journal of the American Planning Association 1994; 60(2), 222-236.

22. Buckle P, Marsh GL, Smale S. Reframing Risk, Hazards, Disasters, and Daily Life: A Report of Research into Local Appreciation of Risks and Threats. International Journal of Mass Emergencies and Disasters 2002; 20(3), 309-324.

23. Stuart-Black J, Coles E, Norman S. Bridging the Divide from Theory to Practice. International Journal of Mass Emergencies and Disasters 2005; 23(3), 177-198.

24. Ender RL, Kim, JCK. (1988). The Design and Implementation of Disaster Mitigation Policy. In Comfort LK (ed.) Managing Disaster: Strategies and Policy Perspectives. Durham, NC: Duke Press; 1988 p69.

25. Kartez JD, Kelley WJ. (1988). Research-based Disaster Planning: Conditions for Implementation. In Comfort LK (ed.) Managing Disaster: Strategies and Policy Perspectives. Durham, NC: Duke Press; 1988 p135.

26. Jackman AM, Beruvides MG. How Much Do Hazard Mitigation Plans Cost? An Analysis of Federal Grant Data. Working paper; submitted 9 July 2012 to the Journal of Emergency Management.

27. Jackman, AM. An Analysis of the Cost of Hazard Mitigation Planning Policy in Local and Regional Government. Ph.D. Thesis. Texas Tech University; 2008.

Chapter 7

Improved Disaster Management Using Data Assimilation

Paul R. Houser[1]

[1]George Mason University, Fairfax, VA, USA

INTRODUCTION

Decision makers must have timely and actionable information to guide their response to emergency situations. For environmental problems, this information is often produced using decision support systems (DSS), which is usually a computer-based, environmental simulation and prediction model that emphasizes access and manipulation of data and algorithms. Using historical time series data, current conditions, and physically-based algorithms the DSS can predict the potential outcomes for various decision scenarios, and may also provide the decision maker with uncertainty and risk estimates. In this way, the

DSS can improve decision making efficiency and accuracy, facilitate decision maker exploration and discovery, communication and information organization, and outreach and education.

An important component of advanced decision support tools is data assimilation. Data assimilation is the application of recursive Bayesian estimation to combine current and past data in an explicit dynamical model, using the model's prognostic equations to provide time continuity and dynamic coupling amongst the fields. Data assimilation aims to utilize both our knowledge of physical processes as embodied in a numerical process model, and information that can be gained from observations, to produce an improved, continuous system state estimate in space and time. When implemented in near-real time, data assimilation can objectively provide decision makers with the timeliest information, as well as provide superior initializations for short term scenario predictions. Data assimilation can also act as a parameter estimation method to help reduce DSS bias and uncertainty.

This chapter will provide an overview of data assimilation theory and its application to decision support tools, and then provide a review of current data assimilation applications in disaster management.

BACKGROUND

Information about environmental conditions is of critical importance to real-world applications such as agricultural production, water resource management, flood prediction, water supply, weather and climate forecasting, and environmental preservation. Improved estimates about current environmental conditions useful for agriculture, ecology, civil engineering, water resources management, rainfall-runoff prediction, atmospheric process studies, climate and weather/climate prediction, and radiation management [1,2].

This information is usually provided to decision makers through Decision Support Systems (DSS). DSS's are generally defined as interactive software-based systems that help to assemble useful information from raw data, documents, knowledge, and models to identify and resolve problems and make decisions. A model-driven DSS emphasizes access to and operation of a statistical, financial, optimization or physical simulation model. A data-driven DSS emphasizes access to and manipulation of a time-series of data and information. Data-driven DSS's combined with analytical model processing provide the highest level of functionality and decision support that is linked to analysis of large collections of historical data.

Physically-based environmental models are often at the heart of powerful DSSs. They rely on a set of well-established physical principles to make current condition assessments and future projections. Physical model simulations are performed on powerful computer platforms, dividing the area of interest into elements in which fluxes and storages are calculated. Environmental parameters are provided by connected databases of observational and calibration data.

Observations are important components of DSSs, providing critical information that mitigates the risk of loss of life and damage to property. Environmental observations are sourced from the numerous disconnected observational networks and systems that have a wide variety of characteristics (Figure 1). Basic monthly, seasonal and annual summaries of temperature, rainfall and other climate elements provide an essential resource for planning endeavors in areas such as agriculture, water resources, emergency management, urban design, insurance, energy supply and demand management and construction.

While ground-based observational networks are improving, the only practical way to observe the environment on continental to global scales is via satellites. Remote sensing can make spatially comprehensive measurements of various components of the environment, but it cannot provide information on the entire system, and the observations represent only an instant in time. Environmental process models may be used to predict the temporal and spatial state variations, but these predictions are often poor, due to model initialization, parameter and forcing, and physics errors. Therefore, an attractive prospect is to combine the strengths of environmental models contained within DSSs and observations and minimize the weaknesses to provide a superior environmental state estimate. This is the goal of data assimilation.

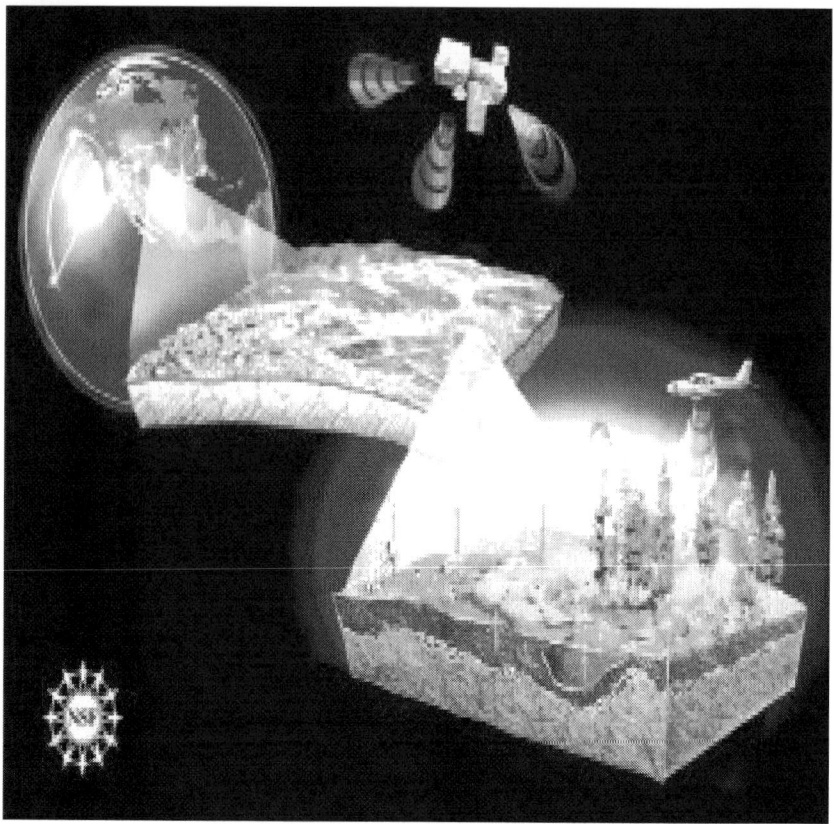

Figure 1: Illustration of an integrated environmental observation network. The network illustrated is the National Science Foundation (NSF) National Ecological Observatory Network (NEON).

Data assimilation combines observations into a dynamical model, using the model's equations to provide time continuity and coupling between the estimated fields. Data assimilation aims to utilize both our environmental process knowledge, as embodied in a numerical computer model, and information that can be gained from observations. Both model predictions and observations are imperfect and we wish to use both synergistically to obtain a more accurate result. Moreover, both contain different kinds of information, that when used together, provide an accuracy level that cannot be obtained individually.

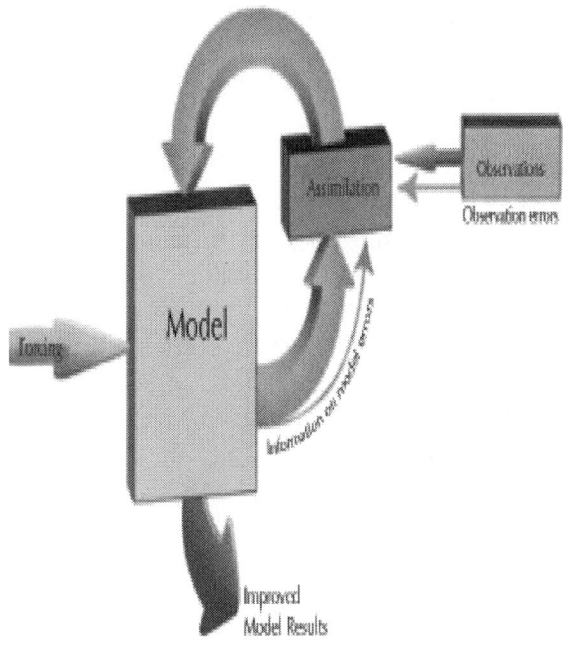

Figure 2: Numerical models contain errors that increase with time due to model imperfections and uncertainties in initial and boundary conditions. Data assimilation minimizes these errors by correcting the model stats using new observations (from http://www.hzg.de/institute/coastal_research/cosyna).

The data assimilation challenge is to merge the spatially comprehensive observations with the dynamically complete but typically poor predictions of an environmental model to yield the best possible system state estimation (Figure 2). In this illustration, the model represents any environmental model that simulates system states. Model biases can be mitigated using a complementary calibration and parameterization process. However, model imperfections will always remain and will be exasperated by uncertain initial and boundary (forcing) conditions. Data assimilation techniques can be used to continuously partially reinitialize the model with information provided by observations. This reinitialization can be constrained by the model physics to assure that it is physically and dynamically realistic. Limited point measurements are often used to calibrate the model(s) and validate the assimilation results [3].

DATA ASSIMILATION

Charney et al. (1969) first suggested combining current and past data in an explicit dynamical model, using the model's prognostic equations to provide time continuity and dynamic coupling amongst the fields [4]. This concept has evolved into a family of techniques known as data assimilation. In essence, data assimilation aims to utilize both our physical process knowledge as embodied in an environmental model, and information that can be gained from observations. Both model predictions and observations are imperfect and we wish to use both synergistically to obtain a more accurate result. Moreover, both contain different kinds of information, that when used together, provide an accuracy level that cannot be obtained when used separately.

Data assimilation techniques were established by meteorologists [5] and have been used very successfully to improve operational weather forecasts. Data assimilation has also been successfully used in oceanography[6] for improving ocean dynamics prediction. Houser et al., (2010) gave an overview of hydrological data assimilation, discussing different data assimilation methods and several case studies in hydrology [7].

Data assimilation was meant for state estimation, but in the broadest sense, data assimilation refers to any use of observational information to improve a model [8]. Basically, there are four methods for "model updating", as follows:

- *Input:* corrects model input forcing errors or replaces model-based forcing with observations, thereby improving the model's predictions;
- *State:* corrects the state or storages of the model so that it comes closer to the observations (state estimation, data assimilation in the narrow sense);
- Parameter: corrects or replaces model parameters with observational information (parameter estimation, calibration);
- *Error correction:* correct the model predictions or state variables by an observed time-integrated error term in order to reduce systematic model bias (e.g. bias correction).

State updating can be justified by lack of knowledge about the model's initial conditions, but with unconstrained state updating, the model logic is foregone, while this is exactly the main strength of dynamic assimilation and modelling. If an intensive update of the state is needed for good results, the model may simply not be able to produce correct state or flux values. In such cases, assimilation for parameter estimation is better advised. The static parameters obtained through off-line calibration, prior to the actual forecast simulations, may not always result in a proper model definition, because of the state and time dependency of parameters or problems in the model structure or input. Often the model validation residuals show the presence of bias, variation in error and a correlation structure.

The data assimilation challenge is: given a (noisy) model of the system dynamics, find the best estimates of system states x^ from (noisy) observations y. Most current approaches to this problem are derived from either the direct observer (i.e., sequential filter) or dynamic observer (i.e., variational through time) techniques (Figure 3).

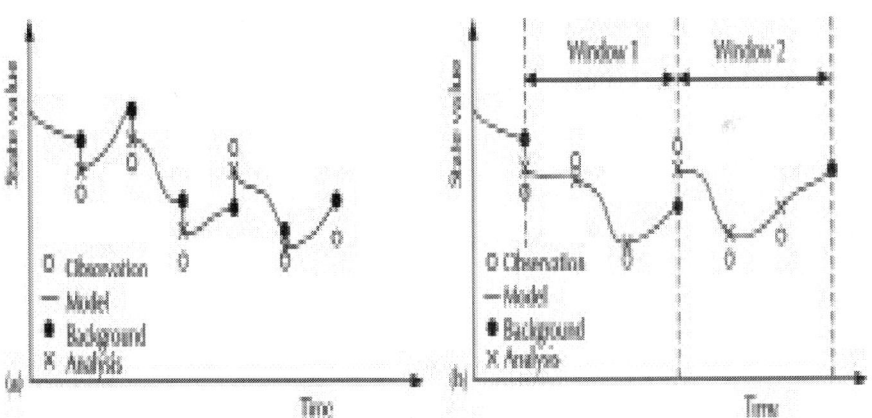

Figure 3: Schematic of the (a) direct observer and (b) dynamic observer assimilation approaches [7].

Direct Observer Assimilation

Direct observer techniques sequentially update the model forecast $\hat{x}_k^b$ (a priori simulation result), using the difference between observation y_k and model predicted observation $\hat{y}_k$, known as the "innovation", whenever observations are available. The predicted observation is calculated from the model predicted or "background" states, indicated by the superscript b. The correction, or analysis increment, added to the background state vector is the innovation multiplied by a weighting factor or gain K. The resulting estimate of the state vector is known as the "analysis", as indicated by the superscript a.

$$\hat{x}_k^a = \hat{x}_k^b + K_k \left(y_k - \hat{y}_k \right) \tag{1}$$

The subscript k refers to the time of the update. For particular assimilation techniques, like the Kalman filter, the gain represents the relative uncertainty in the observation and model variances, and is a number between 0 and 1 in the scalar case. If the uncertainty of the predicted observation (as calculated from the background states and their uncertainty) is large relative to the uncertainty of the actual observation, then the analysis state vector takes on values that will closely yield the actual observation. Conversely, if the uncertainty of the predicted observation is small relative to the uncertainty of the actual observation, then the analysis state vector is unchanged from the original background value. The commonly used direct observer methods are: (i) direct insertion; (ii) statistical correction; (iii) successive correction; (iv) analysis correction; (v) nudging; (vi) optimal interpolation/statistical interpolation; (vii) 3-D variational, 3D-Var; and (viii) Kalman filter and variants [7].

While approaches like direct insertion, nudging and optimal interpolation are computationally efficient and easy to implement, the updates do not account for observation uncertainty or utilize system dynamics in estimating model background state uncertainty, and information on estimation uncertainty is limited. The Kalman filter, while computationally demanding in its pure form, can be adapted for near-real-time application and provides information on estimation uncertainty. However, it has only limited capability to deal with different types of model errors, and necessary linearization approximations can lead to unstable solutions. The Ensemble Kalman filter (EnKF), while it can be computationally demanding (depending on the size of the ensemble) is well suited for near-real-time applications without any need for linearization, is robust, very flexible and easy to use, and is able to accommodate a wide range of model error descriptions.

Direct insertion: One of the earliest and most simplistic approaches to data assimilation is direct insertion. As the name suggests, the forecast model states are directly replaced with the observations by assuming that K = I, the unity matrix. This approach makes the explicit assumption that the model is wrong (has no useful information) and that the observations are right, which both disregards important information provided by the model and preserves observational errors. The risk of this approach is that unbalanced state estimates may result, which causes model shocks: the model will attempt to restore the dynamic balance that would have existed without insertion. A further key disadvantage of this approach is that model physics are solely relied upon to propagate the information to unobserved parts of the system [9,10].

Statistical correction: A derivative of the direct insertion approach is the statistical correction approach, which adjusts the mean and variance of the model states to match those of the observations. This approach assumes the model pattern is correct but contains a non-uniform bias. First, the predicted observations are scaled by the ratio of observational field standard deviation to predicted field standard deviation. Second, the scaled predicted observational field is given a block shift by the difference between the means of the predicted observational field and the observational field [9]. This approach also relies upon the model physics to propagate the information to unobserved parts of the system.

Successive correction: The successive corrections method (SCM) [5,11-13] is also known as observation nudging. The scheme begins with an a priori state estimate (background field) for an individual (scalar) variable, which is successively adjusted by nearby observations in a series of scans (iterations, n) through the data. The advantage of this method lies in its simplicity. However, in case of observational error or different sources (and accuracies) of observations, this scheme is not a good option for assimilation, since information on the observational accuracy is not accounted for. Mostly, this approach assumes that the observations are more accurate than model forecasts, with the observations fitted as closely as is consistent. Furthermore, the radii of influence are user-defined and should be determined by trial and error or more sophisticated methods that reduce the advantage of its simplicity. The weighting functions are empirically chosen and are not derived based on physical or statistical properties. Obviously, this method is not effective in data sparse regions.

Analysis correction: This is a modification to the successive correction approach that is applied consecutively to each observation [14]. In practice, the observation update is mostly neglected and further assumptions make the update equation equivalent to that for optimal interpolation [15].

Nudging: Nudging or Newtonian relaxation consists of adding a term to the prognostic model equations that causes the solution to be gradually relaxed towards the observations. Nudging is very similar to the successive corrections technique and only differs in the fact that through the numerical model the time dimension is included. Two distinct approaches have been developed [16]. In analysis nudging, the nudging term for a given variable is proportional to the difference between the model simulation at a given grid point and an "analysis" of observations (i.e., processed observations) calculated at the corresponding grid point. For observation nudging, the difference between the model simulation and the observed state is calculated at the observation locations.

Optimal interpolation: The optimal interpolation (OI) approach, sometimes referred to as statistical interpolation, is a minimum variance method that is closely related to kriging. OI approximates the "optimal" solution often with a "fixed" structure for all time steps, given by prescribed variances and a correlation function determined only by distance [17]. Sometimes, the variances are allowed to evolve in time, while keeping the correlation structure time-invariant.

3-D variational: This approach directly solves the iterative minimization problem given [18]. The same approximation for the background covariance matrix as in the optimal interpolation approach is typically used.

Kalman filter: The optimal analysis state estimate for linear or linearized systems (Kalman or Extended Kalman filter, EKF) can be found through a linear update equation with a Kalman gain that aims at minimizing the analysis error (co)variance of the analysis state estimate [19]. The essential feature which distinguishes the family of Kalman filter approaches from more static techniques, like optimal interpolation, is the dynamic updating of the forecast (background) error covariance through time. In the traditional Kalman filter (KF) approach this is achieved by application of standard error propagation theory, using a (tangent) linear model. (The only difference between the Kalman filter and the Extended Kalman filter is that the forecast model is linearized using a Taylor series expansion in the latter; the same forecast and update equations are used for each approach.)

A further approach to estimating the state covariance matrix is the Ensemble Kalman filter (EnKF). As the name suggests, the covariances are calculated from an ensemble of state forecasts using the Monte Carlo approach rather than a single discrete forecast of covariances [20].

Dynamic Observer Assimilation

The dynamic observer techniques find the best fit between the forecast model state and the observations, subject to the initial state vector uncertainty and observation uncertainty, by minimizing over space and time an objective or penalty function, including a background and observation penalty term. To minimize the objective function over time, an assimilation time "window" is defined and an "adjoint" model is typically used to find the derivatives of the objective function with respect to the initial model state vector. The adjoint is a mathematical operator that allows one to determine the sensitivity of the objective function to changes in the solution of the state equations by a single forward and backward pass over the assimilation window. While an adjoint is not strictly required (i.e., a number of forward passes can be used to numerically approximate the objective function derivatives with respect to each state), it makes the problem computationally tractable. The dynamic observer techniques can be considered simply as an optimization or calibration problem, where the state vector – not the model parameters – at the beginning of each assimilation window is "calibrated" to the observations over that time period. The dynamic observer techniques can be formulated with: (i) strong constraint (variational); (ii) weak constraint (dual variational or representer methods).

Dynamic observer methods are well suited for smoothing problems, but provide information on estimation accuracy only at considerable computational cost. Moreover, adjoints are not available for many existing environmental models, and the development of robust adjoint models is difficult due to the non-linear nature of environmental processes.

4D-Var: In its pure form, the 4-D (3-D in space, 1-D in time) "variational" (otherwise known as Gauss-Markov) dynamic observer assimilation methods use an adjoint to efficiently compute the derivatives of the objective function with respect to each of the initial state vector values. Solution to the variational problem is then achieved by minimization and iteration. In practical applications the number of iterations is usually constrained to a small number.

Given a model integration with finite time interval, and assuming a perfect model, 4D-Var and the Kalman filter yield the same result at the end of the assimilation time interval. Inside the time interval, 4D-Var is more optimal, because it uses all observations at once (before and after the time step of analysis), i.e., it is a smoother. A disadvantage of sequential methods is the discontinuity in the corrections, which causes model shocks. Through variational methods, there is a larger potential for dynamically based balanced analyses, which will always be situated within the model climatology. Operational 4D-Var assumes a perfect model: no model error can be included. With the inclusion of model error, coupled equations are to be solved for minimization. Through Kalman filtering it is in general simpler to account for model error.

Both the Kalman filter and 3D/4D-Var rely on the validity of the linearity assumption. Adjoints depend on this assumption and incremental 4D-Var is even more sensitive to linearity. Uncertainty estimates via the Hessian are critically dependent on a valid linearization. Furthermore, with variational assimilation it is more difficult to obtain an estimate of the quality of the analysis or of the state's uncertainty after updating. In the framework of estimation theory, the goal of variational assimilation is the estimation of the conditional mode (maximum a posteriori probability) estimate, while for the Kalman filter the conditional mean (minimum variance) estimate is sought.

Hybrid assimilation methods have been explored in which a sequential method is used to produce the a priori state error or background error covariance for variational assimilation [7].

REVIEW OF CURRENT DATA ASSIMILATION APPLICATIONS IN DISASTER MANAGEMENT

Weather Forecasting

During the last three decades, data assimilation has gradually reached a mature center stage position at operational numerical weather prediction centers, and are largely responsible for the significant advances in weather forecast accuracy [21]. Improved weather forecasts are critical for better informing the public and decision makers about impending severe weather events such as tropical storms, tornados, frozen precipitation events, wind hazards, droughts, and flooding.

The basis for improved weather prediction using data assimilation is to improve the initial state, which results in an improved forecast. Initial work was based on hand interpolations that combined present and past observations with model results [22-24]. This tedious procedure was replaced by automatic objective analysis [12, 25-27].

Currently, data assimilation is available and implemented worldwide at operational numerical weather prediction centers. The impact of adopting data assimilation in numerical weather prediction was qualified as a substantial, resulting in an improvement in NWP quality and accuracy [28]. Combined with improvement in error specifications and with a large increase in a variety of observations has led to improvements in NWP accuracy [29].

The development of the global positioning system (GPS) satellites has facilitated the use of radio occultation (RO) techniques for sounding the earth's atmosphere. RO is a remote sensing technique that relies on the detection of a change or refraction in a radio signal as it passes through the atmosphere. The degree of refraction depends on the gradients of density and the water vapor. These global measurements are actually commensurable with radiosonde soundings in accuracy [30]. Assimilations of the RO retrieved data have exhibited promising impact on regional as well as global weather predictions [31,32].

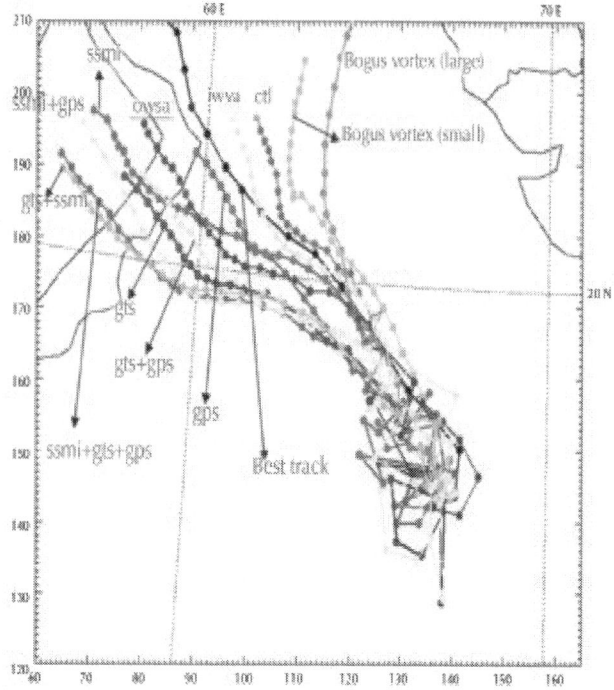

Figure 4: The best track from JTWC (black line) for Cyclone Gonu (2007) and simulated tracks for experiments CTL and GPS without and with the RO data assimilated, respectively, and other experiments with different data, GTS (radiosondes), SSMI (SSM/I retrieved oceanic near-surface wind speed OWS and integrated precipitable water PW, denoted as OWSA an IWVA, respectively), bogus vortex 1 (large vortex) and bogus vortex2 (small vortex) [33].

The impact of GPS radio occultation data assimilation on severe weather predictions was demonstrated in East Asia [33]. These observations were assimilated in the Weather Research and Forecasting (WRF) model's using a three-dimensional variational (3DVAR) data assimilation system to improve the initial analysis of the model. The GPS RO data assimilation may improve prediction of severe weather such as typhoons. These positive impacts are seen not only in typhoon track prediction but also in prediction of local heavy rainfall associated with severe weather over Taiwan. From a successive evaluation of skill scores for real-time forecasts on frontal systems operationally conducted over a longer period and predictions of six typhoons in 2008, assimilation of GPS RO data appears to have some positive impact on regional weather predictions, on top of existent assimilation with all other observations.

The impact of GPS RO assimilation on Tropical Cyclone Gonu (2007) was studied over the western Indian Ocean [33]. Gonu was one of the most intense in regional history and had asevere impact. The positive impact of GPS RO data on track prediction is clearly seen in Figure 4. It is surprisingly found that assimilations with all other data (including SSM/I, GTS and their combinations) do not outperform the run with GTS+GPS or even the run with GPS RO data only.

Flood Management: Early Warning, Monitoring, and Damage Assessments

Flood forecasting using numerical models and data assimilation techniques provides extended lead-time and improved accuracy for flood information useful for residents, local authorities and emergency services. The use of data assimilation in operational hydrologic forecasting predates its use in weather forecasting and oceanography. Examples include updating of snow model states and the use of observed streamflow to make short-term adjustments to the simulated streamflow. However, despite its early adoption, more advanced methods of data assimilation (i.e. Kalman filtering) has yet to take firm root in operational hydrologic forecasting.

Operational hydrologic data assimilation typically uses telemetered, near real-time measurements of river levels and flows, and raingauge or Doppler radar precipitation estimates as inputs to a computer-based flood forecasting system. Model outputs include minute to several day forecasts for river levels, river flows, and reservoir and lake levels. Forecasts can be extended further using weather forecasts, and can include snowmelt processes and river control operations.

Hydrologic rainfall-runoff models are used to estimate river flows from rainfall observations and forecasts. These models may take into account local catchment topography, soils, vegetation, temperature, river flow hydrodynamics, and structure operations and backwater effects. These models are enhanced using data assimilation methods such as error prediction, state updating, and parameter updating techniques. Forecast uncertainties can arise and propagate through the modeling network from errors in model parameters, initial conditions, boundary conditions, data inputs and model physics.

As part of the European Union near real time flood forecasting, warning and management "FloodMan" project data-assimilation techniques were developed, demonstrated and validated for integrated hydrological and hydraulic models in a pilot study of the Rhine River [34]. The model combines a hydraulic model (SOBEK) representing the Rhine River between Andernach and Düsseldorf with a hydrological model (HBV) for the Sieg tributary. To increase the accuracy of flood forecasts, data assimilation is applied, using measured water levels at Bonn and Cologne to adapt bed roughness and lateral discharges until the calculated water levels agree with the measured water levels.

The data-assimilation technique computes model corrections based on the assumption that the uncertainties in model output and observations are known and are normally distributed. The data-assimilation algorithm compares the observed value with the calculated value, and makes corrections to the parameters. The changes are made taking into consideration the uncertainty of the parameters and measured data. For example, a mean water level measurement is accurate compared to calculated water levels. Therefore the water level calculated with data-assimilation will be closer to the measured water level than the calculated without data-assimilation.

The data-assimilation algorithm is applied to the flood of December 1993 (Figure 5). The data-assimilation improves the water level forecast. The small differences between forecast and measured data are due to the perfect forecast for the input at Andernach. The adaptations by the data-assimilation on the model parameters were small indication a well calibrated hydraulic model for 1993 flood and that robust data assimilation procedure.

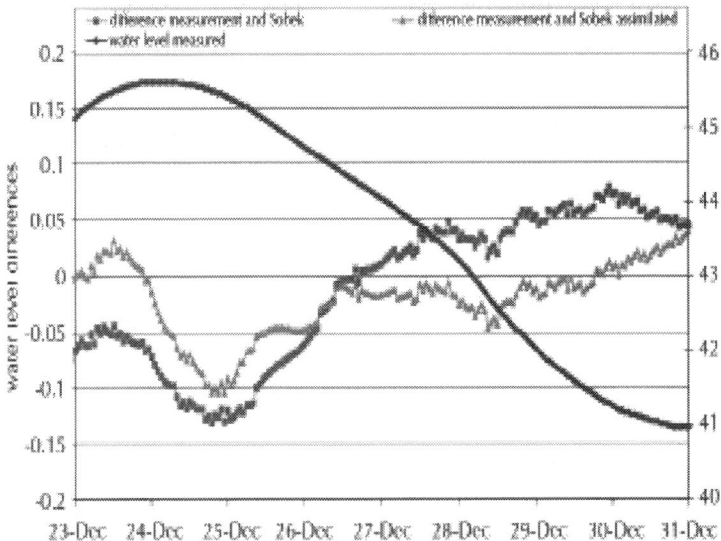

Figure 5: Streamflow forecasts using data assimilation. The blue line depicts the difference between forecast and measured water levels without data-assimilation, the red line the differences with data-assimilation [34].

Drought Management

Droughts are environmental disasters that occur in virtually all climates, and are generally related to reduced precipitation for an extended period of time. High temperature, high wind, low humidity; rainfall timing, intensity and duration, also play a significant role in droughts [35]. Aridity is a permanent feature of climate related to low rainfall areas [36], while drought is a temporary anomaly, lasting from months to several years. Population growth, agricultural and industrial expansion, energy demands for water, climate change, and water contamination further amplify the effects of drought and water scarcity.

Droughts impact both surface and groundwater, leading to reduced water supply and quality, crop failure, reduced livestock range, reduced power supply, disturbed riparian habitat, and deferred recreation [37]. Therefore, droughts are of great importance in the planning and management of water resources.

Droughts rank first among all natural hazards when measured in terms of the number of people affected [38]. Hazard events were ranked based on the degree of severity, the length of event, total areal extent, total loss of life, total economic loss, social effect, long-term impact, suddenness, and occurrence of associated hazards [39]. It was found that drought stood first based on most of the hazard characteristics. Other natural hazards, which followed droughts in terms of their rank, are tropical cyclones, regional floods, earthquakes, and volcanoes.

The Gravity Recovery and Climate Experiment (GRACE) satellite mission, launched in 2002 measures monthly changes in total water storage over large areas, which can help to assess change in water supply on and beneath the land surface. However, the coarse spatial and temporal resolutions of GRACE, and its lack of information on the vertical distribution of the observed mass changes limits its utility unless it is combined with other sources of information. In order to increase the resolution, eliminate the time lag, and isolate groundwater and other components from total terrestrial water storage, the GRACE data was integrated with other ground- and space-based meteorological observations (precipitation, solar radiation, etc.) within the Catchment Land Surface Model, using Ensemble Kalman smoother type data assimilation [40]. The resulting fields of soil moisture and groundwater storage variations are then used to generate drought indicators based on the cumulative distribution function of wetness conditions during 1948-2009 simulated by the Catchment model [41] (Figure 6).

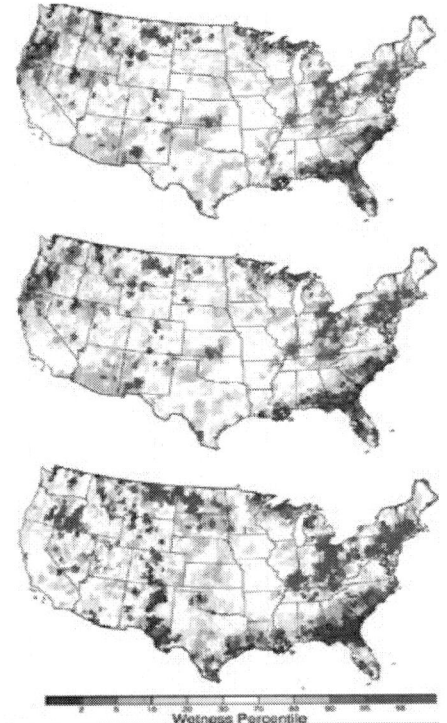

Figure 6: GRACE data assimilation based drought indicators for surface (top) and root zone (middle) soil moisture and groundwater (bottom) for 26 December 2011, expressed as percentiles relative to conditions during the 1948-2011 simulated record [41].

There are several aquifers in the U.S. that have been depleted in that way over the past century, such as the southern half of the High Plains aquifer in the central U.S. If the groundwater drought indicator map accounted for human-induced depletion, such regions would be red all the time, which would not be useful for evaluating current wetness conditions relative to previous conditions. On time scales of weeks to ten years, we expect that these maps will be reasonably well correlated with measured water table variations over spatial scales of 25 km (16 miles) or more. However, users should not assume a direct correspondence between these groundwater percentiles and measured groundwater levels over multiple decades. The color-coded maps show how much water is stored now as a probability of occurrence in the record from 1948 to the present.

Radiation Guidance and Monitoring

Decision makers must have the information needed to react in a rapid and appropriate manner before, during and immediately after an accidental or intentional contamination of the environment. Decision support systems are needed to estimate the likely evolution of the environmental contamination. The primary goal is to determine the area likely to be affected by a possible release and to obtain an estimate of the potential maximum environmental consequences. In the early phases of an accident the main goal is to provide a forecast of the magnitude and geographical coverage of the potential environmental consequences. It is important to know the prevailing and forecasted meteorological conditions in the local area. Also the status of the source should be known in detail. Depending on the meteorological situation and the model used, trajectories may be calculated first to give a rough estimation of the plume transport.

Dispersion models driven with weather data and best-estimate source information can be used. When results of radiological measurements are available they can be used to improve model calculations by data assimilation. Atmospheric dispersion modeling of radioactive material in radiological risk assessment and emergency response has evolved significantly over the past 50 years. The three types of dispersion models are the Gaussian plume, Lagrangian-puff and particle random walk, and computational fluid dynamic models. When data from radiological measurements are available, they should be taken into account in the consequence assessment and used to correct and update model calculation results (data assimilation). Because observations are often sparse in emergency situations, data assimilation procedures should be designed to handle cases with only a few measurements. Even simple dispersion models would benefit from data assimilation, and may also run faster to provide critical time-sensitive information to decision makers [42].

Rojas-Palma et al., (2005) describe an in-depth effort to integrate a suite of computer codes, with different degrees of complexity, into a European real-time, on-line decision support system for off-site management of nuclear emergencies (the RODOS system) [42]. The resulting modeling system describes the transport and dispersion of radionuclides in both atmospheric and aquatic systems, as well as their impact on the food chain.

RODOS predicts the values of many quantities that are likely to be of interest to decision makers after an accident (e.g. activity concentration in air, deposition, concentration in foods, external dose rates, concentrations in water bodies). The predictions will not exactly reflect the situation after an accident, as the models use a number of assumptions that are appropriate to the average situation across large areas of Europe, rather than to the particular conditions of the area affected by the accident. In the period immediately after the accident there will be a limited amount of information available from monitoring programs. To make the best use of this information, it is necessary to correct the RODOS predictions in light of the available measurements.

A 2 km RODOS test case was generated with a release point at Risø, Denmark, and 23 detector points surrounding Risø (Figure 7,8). The meteorological situation was a 7m/s westerly wind at 60 m above ground with neutral stability, and no rain.

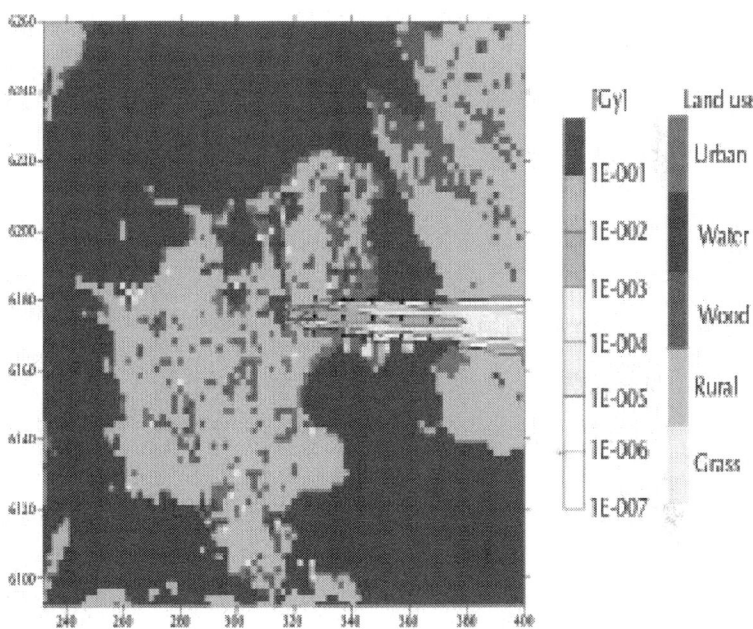

Figure 7: Gamma dose field from puffs in measurement generating run. The 15 detector points 10 to 50 km east of the release point are seen as black squares, the 8 points surrounding the release point are not marked. The background picture is the land use map [42].

In general, in off-site emergency management, data assimilation will prove useful throughout the different stages of the accident. In the assessment of the consequences during the early phase, in the improvement of prior assumptions based solely on expert judgement, and when there is a clear need for longer-term predictions to assess the radiological impact on the food chain.

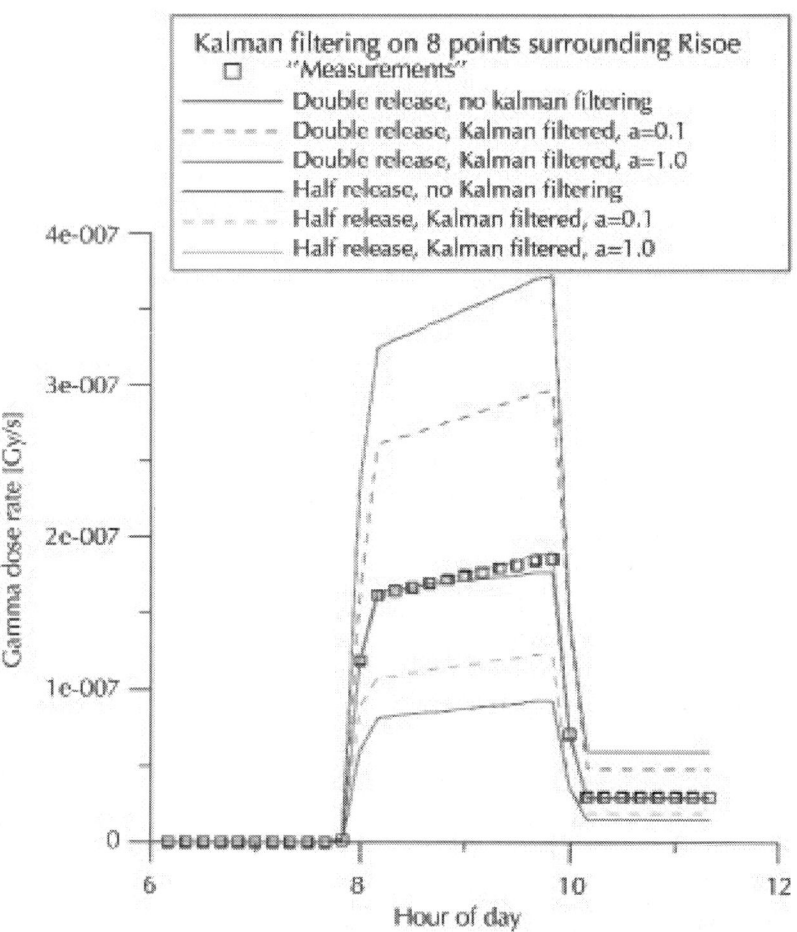

Figure 8: Gamma dose rates 50 km from the release point, right below the plume. Kalman filtering applied in runs with the release doubled and halved relative to that having generated the "measurements", and with different release rate uncertainties "a" [42].

Tsunami Warnings and Forecasts

A tsunami is a series of waves that can move on shore rapidly, but last for several hours and flood coastal communities with little warning. Tsunamis can be triggered by a variety of geological processes such as earthquakes, landslides, volcanic eruptions, or meteorite impacts. Throughout history, Tsunami's have taken many lives in coastal regions around the world. In the wake of the catastrophic 2004 Indian Ocean tsunami, which caused over 200,000 deaths and widespread destruction, many governmental organizations have increased their efforts to diminish the potential impacts of a tsunami by strengthening tsunami detection, warning, education and preparedness efforts [43].

In contrast to forecasting other natural hazards such as hurricanes or floods, near-real-time tsunami forecast models must produce predictions after a seismic event has been detected, but before the event arrive at the coast. These forecasts provide emergency managers near-realtime information about the time of first impact as well as the sizes and duration of the tsunami waves, and give an estimate of the area of inundation. The entire forecasting process has to be completed very quickly, to allow time for evacuation efforts. The entire forecast, including data acquisition, data assimilation and inundation projections, must take place within a few hours [44].

Titov et al., (2003) presented a method for tsunami forecasting that combining real-time data from tsunameters with numerical model estimates to provide site- and event-specific forecasts for tsunamis in real time [45]. Observational networks will never be sufficiently dense because the ocean is vast. Establishing and maintaining monitoring stations is costly and difficult, especially in deep water. Numerical model accuracy is inherently limited by errors in bathymetry and topography and uncertainties in the generating mechanism. But combined, these techniques can provide reliable tsunami forecasts, as is demonstrated in the Short-term Inundation Forecasting (SIFT) system. The Method of Splitting Tsunamis (MOST) numerical model is run in two steps or modes. In the data assimilation mode, the model is adjusted "on-the-fly" by a real-time data stream to provide the best fit to the data. In the forecast mode, the model uses the simulation scenario obtained in the first step and extends the simulation to locations where measured data is not available, providing the forecast. An effective implementation of the inversion is achieved by using a discrete set of Green's functions to form a model source. The algorithm chooses the best fit to a given tsunameter data among a limited number of unit solution combinations by direct sorting, using a choice of misfit functions (Figure 9).

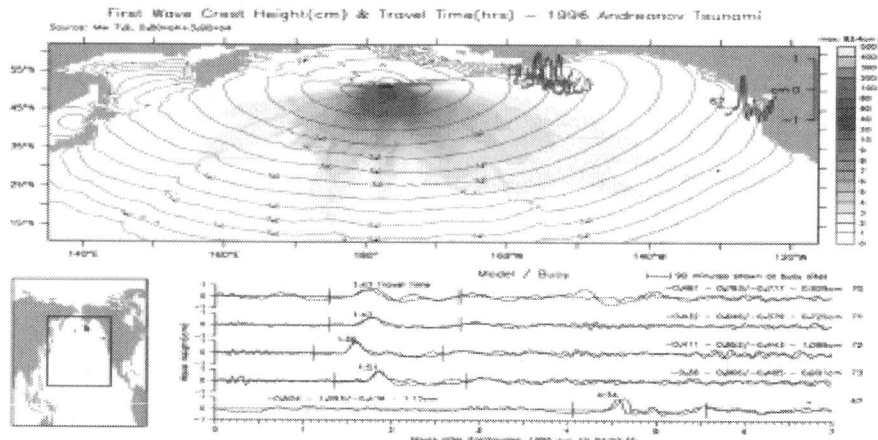

Figure 9: Results of MOST data assimilation for 1996 Anderanov Island tsunami [45]. Top frame shows the source inferred by the data assimilation (black rectangles show unit sources' fault plains), maximum computed amplitudes of tsunami from this source (filled colored contours), travel time contours in hours after earthquake (solid lines), and locations of the bottom pressure recorders. Bottom frame shows a reference map (left) and comparison of the model (blue) and bottom pressure recorder data (magenta).

CONCLUSIONS

This chapter provided an overview of data assimilation theory and its application to decision support tools, and provided 5 examples of operational data assimilation applications in disaster management. These included tsunami warning, radiation guidance and monitoring, flood and drought management, and weather forecasting.

Information about environmental conditions is of critical importance to real-world applications disaster management in areas such as agricultural production, water resource management, flood prediction, water supply, weather and climate forecasting, and environmental preservation. This information is usually provided to decision makers through Decision Support Systems (DSS). Observations are important components of DSSs, providing critical information that mitigate the risk of loss of life and damage to property. Environmental process models are used in DSSs to predict the temporal and spatial state variations, but these predictions are often poor, due to model initialization, parameter and forcing, and physics errors. Therefore, we must combine the strengths of environmental models contained within DSSs and observations and minimize the weaknesses to provide a superior environmental state estimate – data assimilation.

Data assimilation merges the spatially comprehensive observations with the dynamically complete but typically poor predictions of an environmental model to yield the best possible system state estimation. Data assimilation aims to utilize both our knowledge of physical processes as embodied in a numerical process model, and information that can be gained from observations, to produce an improved, continuous system state estimate in space and time. When implemented in near-real time, data assimilation can objectively provide decision makers with the timeliest information, as well as provide superior initializations for short term scenario predictions. Data assimilation can also act as a parameter estimation method to help reduce DSS bias and uncertainty.

REFERENCES

1.   P Houser, 2002Air and Water Monitoring for Homeland Security, Earth Observation Magazine, 1183336

2.   P Houser, M. F Hutchinson, P Viterbo, Hervé Douville, J., and Running, S. W. 2004Terrestrial data assimilation, Chapter C.4 in Vegetation, Water, Humans and the Climate. Global Change-The IGB Series. Kabat, P. et al. (eds). Springer, Berlin.

3.   J. P Walker, and P. R Houser, 2005Hydrologic Data Assimilation, In: A. Aswathanarayana (Ed.), Advances in Water Science Methodologies, A.A. Balkema, TheNetherlands, 230pp.

4.   J. G Charney, M Halem, and R Jastrow, 1969Use of incomplete historical data to infer the present state of the atmosphere. J. Atmos. Sci., 2611601163

5.   R Daley, 1991Atmospheric data analysis. Cambridge University Press, 460 pp.

6.   A. F Bennett, 1992Inverse methods in physical oceanography. Cambridge University Press, 346 pp.

7.   P. R Houser, G De Lannoy, and J. P Walker, Land Surface Data Assimilation, p549598In: Lahoz, W., Khatattov, B. and Menard, R. (Eds.), Data Assimilation: Making sense of observations, Springer, The Netherlands, 2010pp.

8.   WMO1992Simulated real-time intercomparison of hydrological models (Tech. Rep. 38Geneva.

9.   P. R Houser, W. J Shuttleworth, J. S Famiglietti, H. V Gupta, K. H Syed, and D. C Goodrich, 1998Integration of soil moisture remote sensing and hydrologic modeling using data assimilation. Water Resour. Res., 3434053420

10.   [10] J. P Walker, G. R Willgoose, and J. D Kalma, 2001One-dimensional soil moisture profile retrieval by assimilation of near-surface observations: A comparison of retrieval algorithms. Adv. Water Resour., 24631650

11.   P Bergthorsson, and B Döös, 1955Numerical weather map analysis. Tellus, 7329340

12.   G. P Cressman, 1959An operational objective analysis system. Mon. Weather Rev., 87367374

13.   A. M Bratseth, 1986Statistical interpolation by means of successive corrections. Tellus, 38A, 439447

14.   A. C Lorenc, R. S Bell, and B Macpherson, 1991The Meteorological Office analysis correction data assimilation scheme. Q. J. R. Meteorol. Soc., 1175989

15.   N. K Nichols, 2001State estimation using measured data in dynamic system models, Lecture notes for the Oxford/RAL Spring School in Quantitative Earth Observation.

16.   D. R Stauffer, and N. L Seaman, 1990Use of four-dimensional data assimilation in a limited-area mesoscale model. Part I: Experiments with synoptic-scale data. Mon. Weather Rev., 11812501277

17.   A Lorenc, 1981A global three-dimensional multivariate statistical interpolation scheme. Mon. Weather Rev., 109701721

18. D Parrish, and J Derber, 1992The National Meteorological Center's spectral statistical interpolation analysis system. Mon. Weather Rev., 12017471763

19. R. E Kalman, 1960A new approach to linear filtering and prediction problems. Trans. ASME, Ser. D, J. Basic Eng., 823545

20. M. R. J Turner, J. P Walker, and P. R Oke, 2007Ensemble Member Generation for Sequential Data Assimilation. Remote Sensing of Environment, 112, doi:10.1016/j.rse.2007.02.042.

21. I Navon, Data assimilation for numerical weather prediction: A Review; Springer: Berlin, Germany, 2009

22. L. F Richardson, 1922Weather prediction by numerical processes. Cambridge University Press. Reprinted by Dover (1965, New York).With a New Introduction by Sydney Chapman, xvi+236

23. J. G Charney, et al1950Numerical integration of the barotropic vorticity equation. Tellus 2237254

24. E Kalnay, 2003Atmospheric modeling, data assimilation and predictability. Cambridge Univ Press, Cambridge, 341

25. H. A Panofsky, 1949Objective weather-map analysis. J Appl Meteor 6386392

26. B Gilchrist, G. P Cressman, 1954An experiment in objective analysis. Tellus 6309318

27. S. L Barnes, 1964A technique for maximizing details in numerical weather map analysis. J Appl Meteor 3395409

28. F Rabier, 2005Overview of global data assimilation developments in numerical weather prediction centers. Q J R Meteorol Soc 1316132153233

29. A. J Simmons, A Hollingsworth, 2002Some aspects of the improvement in skill of numerical weather prediction. Q J R Meteorol Soc 128647677

30. E. R Kursinski, S. S Leroy, and B Herman, 2000The GPS radio occultation technique. Terr. Atmos. Oceanic Sci., 1153114

31. Y Kuo, H. , X Zou, W Huang, 1997The impact of GPS data on the prediction of an extratropical cyclone: An observing system simulation experiment. J. Dyn. Atmos. Ocean., 27439470

32. S-Y Chen, C-Y Huang, Y-H Kuo, Y-R Guo, S Sokolovskiy, 2009Typhoon predictions with GPS radio occultation data

assimilations by WRF-VAR using local and nonlocal operators. Terr Atmos Oceanic Sci 20133154

33. C Huang, Y. , Y. -H Kuo, S. -Y Chen, C. -T Terng, F. -C Chien, P. -L Lin, M. -T Kueh, S. -H Chen, M. -J Yang, C. -J Wang, and A. S. K. A. V. P Rao, 2010Impact of GPS radio occultation data assimilation on regional weather predictions. GPS Solutions, 143549DOI:s10291-009-0144-1.

34. C. J. M Vermeulen, H. J Barneveld, H. J Huizinga, F. J Havinga, 2005Data-assimilation in flood forecasting using time series and satellite data. International conference on innovation advances and implementation of flood forecasting technology, 1719October 2005, Tromsø, Norway

35. A. K Mishra, V. P Singh, 2010A review of drought concepts. J Hydrol 39120221

36. D. A Wilhite, 1992Preparing for Drought: A Guidebook for Developing Countries, Climate Unit, United Nations Environment Program, Nairobi, Kenya.

37. [37]W. E Riebsame, S. A Changnon, T. R Karl, 1991Drought and Natural Resource Management in the United States: Impacts and Implications of the 1987-1989 Drought. Westview Press, Boulder, CO, 174

38. D. A Wilhite, Drought as a natural hazard: concepts and definitions. D.A. Wilhite (Ed.), Drought: A Global Assessment, 1RoutledgeNew York (2000118

39. E. A Bryant, Natural Hazards, Cambridge University Press, Cambridge (1991

40. B. F Zaitchik, M Rodell, R. H Reichle, 2008Assimilation of GRACE terrestrial water storage data into a land surface model: results for the Mississippi river basin. J. Hydrometeorol. 9535548

41. R Houborg, M Rodell, B Li, R Reichle, and B Zaitchik, Drought indicators based on model assimilated GRACE terrestrial water storage observations, Wat. Resour. Res., 48, W07525, doi:10.1029/2011WR011291,2012

42. C Rojas-palma, Data assimilation for off site nuclear emergency management," Tech. Rep., SCK-CEN, DAONEM final report, RODOS(RA5)-RE(04)-01, 2005

43. NRC2011Tsunami Warning and Preparedness: An Assessment of the US Tsunami Program and the Nation's Preparedness Efforts. National Academies Press, Washington, DC, 284 pp.

44. P. M Whitmore, H Benz, M Bolton, G. L Crawford, L Dengler, G Fryer, J Goltz, R Hansen, K Kryzanowski, S Malone, D Oppenheimer, E Petty, G Rogers, and J Wilson, 2008NOAA/ WestCoast and Alaska Tsunami Warning Center Pacific Ocean response criteria. Science of Tsunami Hazards 272121

45. V. V Titov, F. I Gonz, E. N Alez, M. C Bernard, H. O Eble, J. C Mofjeld, A. J Newman, Venturato (2004Real-time tsunami forecasting: Challenges and solutions. Nat. Haz., 35(1), 41-58.

Citations

CHAPTER 1

Sunghwan Lee, Sangun Park, and Wooju Kim, "The Importance of Social Value in the Evaluation of Web Services in the Public Sector," International Journal of Distributed Sensor Networks, Article ID 459804, in press.

CHAPTER 2

C. Emdad Haque and M. Salim Uddin (2013). Disaster Management Discourse in Bangladesh: A Shift from Post-Event Response to the Preparedness and Mitigation Approach Through Institutional Partnerships, Approaches to Disaster Management - Examining the Implications of Hazards, Emergencies and Disasters, Prof. John Tiefenbacher (Ed.), ISBN: 978-953-51-1093-4, InTech, DOI: 10.5772/54973.

CHAPTER 3

Roxana L. Ciurean, Dagmar Schröter and Thomas Glade (2013). Conceptual Frameworks of Vulnerability Assessments for Natural Disasters Reduction, Approaches to Disaster Management - Examining the Implications of Hazards, Emergencies and Disasters, Prof. John Tiefenbacher (Ed.), ISBN: 978-953-51-1093-4, InTech, DOI: 10.5772/55538.

CHAPTER 4

Sima Ajami (2013). The Role of Earthquake Information Management System to Reduce Destruction in Disasters with Earthquake Approach, Approaches to Disaster Management - Examining the Implications of Hazards, Emergencies and Disasters, Prof. John Tiefenbacher (Ed.), ISBN: 978-953-51-1093-4, InTech, DOI: 10.5772/53612.

CHAPTER 5

Diane Brand and Hugh Nicholson (2013). Learning from Lisbon: Contemporary Cities in the Aftermath of Natural Disasters, Approaches to Disaster Management - Examining the Implications of Hazards, Emergencies and Disasters, Prof. John Tiefenbacher (Ed.), ISBN: 978-953-51-1093-4, InTech, DOI: 10.5772/53635.

CHAPTER 6

Andrea M. Jackman and Mario G. Beruvides (2013). Hazard Mitigation Planning in the United States: Historical Perspectives, Cultural Influences, and Current Challenges, Approaches to Disaster Management - Examining the Implications of Hazards, Emergencies and Disasters, Prof. John Tiefenbacher (Ed.), ISBN: 978-953-51-1093-4, InTech, DOI: 10.5772/54209.

CHAPTER 7

Paul R. Houser (2013). Improved Disaster Management Using Data Assimilation, Approaches to Disaster Management - Examining the Implications of Hazards, Emergencies and Disasters, Prof. John Tiefenbacher (Ed.), ISBN: 978-953-51-1093-4, InTech, DOI: 10.5772/55840.

Index